α_2-Adrenergic Receptors

α_2-Adrenergic Receptors
Structure, Function and Therapeutic Implications

Edited by

Stephen M. Lanier and **Lee E. Limbird**

Medical University of South Carolina
USA

and

Vanderbilt University Medical Center
USA

harwood academic publishers
Australia • Canada • China • France • Germany • India • Japan
Luxembourg • Malaysia • The Netherlands • Russia • Singapore
Switzerland • Thailand • United Kingdom

Copyright © 1997 OPA (Overseas Publishers Association) Amsterdam B.V. Published in The Netherlands by Harwood Academic Publishers.

All rights reserved.

No part of this book may be reproduced or utilized in any form or by any means, electronic or mechanical, including photocopying and recording, or by any information storage or retrieval system, without permission in writing from the publisher. Printed in Singapore.

Amsteldijk 166
1st Floor
1079 LH Amsterdam
The Netherlands

British Library Cataloguing in Publication Data

α_2-adrenergic receptors: structure, function and
   therapeutic implications
    1. α-adrenoceptors 2. Cardiovascular receptors
    3. Cardiovascular pharmacology 4. α-adrenoceptors –
   Congresses 5. Cardiovascular receptors – Congresses
   6. Cardiovascular pharmacology – Congresses
   I. Lanier, Stephen M. II. Limbird, Lee E.
   615.7'1

ISBN   90-5702-019-X

CONTENTS

Preface		vii
Contributors		ix
Acknowledgements		xvii
Meeting Summary		xix
	Identification, Characterization and Subclassification of α_2-Adrenoceptors: An Overview *J. Paul Hieble, Robert R. Ruffolo, Jr.* and *Klaus Starke*	1
1	α_{2a} Adrenergic Receptor Peptides and G Proteins: Structure and Mechanism *Richard R. Neubig, Joan M. Taylor, Susan M. Wade* and *Shing-Zhao Yang*	19
2	Delineating Ligand-specific Structural Changes in Adrenergic Receptors by Use of Fluorescence Spectroscopy *Ulrik Gether, Sansan Lin* and *Brian K. Kobilka*	31
3	Characterization of Stereoselective Interactions of Catecholamines with the α_{2a}-Adrenoceptor *via* Site Directed Mutagenesis *J. Paul Hieble, Andreas Hehr, Ying-Ou Li, Diane P. Naselsky* and *Robert R. Ruffolo, Jr.*	43
4	Regulating Signal Transfer from Receptor to G-protein *Motohiko Sato, Guangyu Wu* and *Stephen M. Lanier*	53
5	Caveolin, an Integral Membrane Protein Component of Caveolae Membranes *In Vivo*: Implications for Signal Transduction *Michael P. Lisanti, Shengwen Li, ZhaoLan Tang, Kenneth S. Song, Erik Kubler* and *Philipp E. Scherer*	63
6	Molecular Determinants of Agonist-Independent Activity in Adrenergic Receptors *Philippe Samama* and *R.J. Lefkowitz*	77
7	α_2-Adrenergic Receptor Activation and Intracellular Ca^{2+} *Karl E. O. Åkerman, Carina I. Holmberg, Hui-Fang Gee, Annika Renvaktar, Sanna Soini, Jyrki P. Kukkonen, Johnny Näsman, Kristian Enkvist, Michael J. Courtney* and *Christian Jansson*	85
8	The Interface Between α_2-Adrenoceptors and Tyrosine Kinase Signalling Pathways *Graeme Milligan, Andrew R. Burt, Moira Wilson* and *Neil G. Anderson*	95
9	Differential Regulation of α_2-Adrenergic Receptor Subtypes by Norepinephrine and Buffers *D. Roselyn Cerutis, Jean D. Deupree, Donald A. Heck, Si-Jia Zhu, Myron L. Toews* and *David B. Bylund*	103
10	Molecular Basis of α_2-Adrenergic Receptor Subtype Regulation by Agonist *Stephen B. Liggett*	113

CONTENTS

11 α_2-Adrenergic Receptor Subtypes Display Different Trafficking Itineraries in Madin-Darby Canine Kidney II Cells
Magdalena Wozniak and *Lee E. Limbird* — 123

12 Intracellular α_{2a}-Adrenergic Receptors in Neurons and GT1 Neurosecretory Cells
Amy Lee and *Kevin R. Lynch* — 129

13 Immunolocalization of Native α_2-Adrenergic Receptor Subtypes: Differential Tissue and Subcellular Localization
Jeremy G. Richman, Yi Huang and *John W. Regan* — 141

14 α_2-Adrenergic Receptors in the Pathophysiology and Treatment of Depression
Husseini K. Manji, Mark E. Schmidt, Fred Grossman, Karon Dawkins and *William Z. Potter* — 149

15 Clinical Applications of α_2-Adrenergic Agonists in the Perioperative Period – Neurobiologic Considerations
Mervyn Maze and *Toshiki Mizobe* — 161

16 α_2 and α_1-Adrenergic Receptors in the Regulation of Peripheral Vascular Function
Michael T. Piascik, Marta S. Smith, Stephanie E. Edelmann, Leigh B. Macmillan and *Lee E. Limbird* — 171

17 Therapeutic Use of α_2-Adrenoceptor Agonists in Glaucoma
J. Burke, C. Manlapaz, A. Kharlamb, E. Runde, E. Padillo, C. Spada, A. Nieves, S. Munk, T. Macdonald, M. Garst, A. Rosenthal, A. Batoosingh, R. David, J. Walt and *L. Wheeler* — 179

INDEX — 189

ASPET COLLOQUIA — 195

PREFACE

This book is a collection of manuscripts prepared by invited participants in the Colloquium *Alpha$_2$-Adrenergic Receptors: Structure, Function and Therapeutic Implications* held in Nashville, TN USA October 25–27, 1995 and sponsored by the American Society of Pharmacology and Experimental Therapeutics.

THE SEVEN SPAN MOTIF

DANCES WITHIN THE MEMBRANE

CELLS SENSE AND RESPOND.

Utilizing the α_2-adrenergic receptor family as a template for G-protein coupled receptors in general, this conference allowed one to explore the receptor molecule from various perspectives including therapeutic initiatives. Several themes important for G-protein coupled receptor signalling permeated the conference. In addition to the dissection of agonist-induced conformational changes in the receptor protein, there was a recurring effort to divide the receptor into functional subdomains that are important for signal transfer and trafficking. In addition to the individual chapters, we have included the Meeting Summary that was published separately in the *Journal of Pharmacology and Experimental Therapeutics*. Also included is a review of the field "Identification, characterization and subclassification of α_2-adrenergic receptors" kindly contributed by Drs. J. Paul Hieble, Robert Ruffolo (SmithKline Beecham Pharmaceuticals) and Klaus Starke (Albert Ludwigs Universität Freiburg). We deeply appreciate their enthusiasm and commitment to the task.

CONTRIBUTORS

Åkerman, K E O
Department of Biochemistry and Pharmacy
ABO Akademi University
Biocity, P O Box 66
FIN-20521, Turku
Finland

Anderson, N G
Hannah Research Institute
AYR, KA6 5HL
UK

Batoosingh, A
Department of Clinical Research
Allergan Inc
Irvine, CA 92715
USA

Benovic, J L
Department of Pharmacology
Thomas Jefferson University
233 S. 10th Street
Philadelphia, PA 19107-5566
USA

Burke, J
Department of Biological Sciences
Allergan Inc
Irvine, CA 92715
USA

Burt, A R
Molecular Pharmacology Group
University of Glasgow
Glasgow, G12 8QQ
UK

Bylund, D B
Department of Pharmacology
University of Nebraska Medical Center
600 South 42nd Street
Omaha, NE 68198-6260
USA

Cerutis, D R
Department of Pharmacology
University of Nebraska Medical Center
600 South 42nd Street
Omaha, NE 68198-6260
USA

Courtney, M J
Department of Biochemistry and Pharmacy
Abo Akademi University
Biocity, P O Box 66
FIN-20521, Turku
Finland

David, R
Department of Clinical Research
Allergan Inc
Irvine, CA 92715
USA

Dawkins, K
Schizoprenia and Mood Disorders
  Program
WSU School of Medicine
UHC 9B
4201 St Antoine Boulevard
Detroit, MI 48201
USA

Dupree, J D
Department of Pharmacology
University of Nebraska Medical Center
600 South 42nd Street
Omaha, NE 68198-6260
USA

Edelmann, S E
Department of Pharmacology
Vascular Biology Research Group
University of Kentucky
College of Medicine
Lexington, KY 40536
USA

Enkvist, K
Department of Biochemistry and
  Pharmacy
Abo Akademi University
Biocity, P O Box 66
FIN-20521, Turku
Finland

Garst, M
Department of Chemical Sciences
Allergan Inc
Irvine, CA 92715
USA

Gee, H-F
Department of Biochemistry and
  Pharmacy
Abo Akademi University
Biocity, P O Box 66
FIN-20521, Turku
Finland

Gether, U
Department of Cardiology
Stanford University Medical Center
157 Beckman Center
Stanford, CA 94305
USA

Grossman, F
Schizoprenia and Mood Disorders
  Program
WSU School of Medicine
UHC 9B
4201 St Antoine Boulevard
Detroit, MI 48201
USA

Heck, D A
Department of Pharmacology
University of Nebraska Medical Center
600 South 42nd Street
Omaha, NE 68198-6260
USA

Hehr, A
Division of Pharmacological Sciences
SmithKline Beecham Pharmaceuticals
709 Swedeland Road
King of Prussia, PA 19406
USA

Hieble, J P
Pharmacological Sciences UW2510
SmithKline Beecham Pharmaceuticals
709 Swedeland Road
P O Box 1539
King of Prussia, PA 19406-0939
USA

Holmberg, C I
Department of Biochemistry and
  Pharmacy
Abo Akademi University
Biocity, P O Box 66
FIN-20521, Turku
Finland

Huang, Y
Department of Physiology
College of Pharmacy
University of Arizona
Tucson, AZ 85721
USA

Jansson, C
Department of Biochemistry and
  Pharmacy
Abo Akademi University
Biocity, P O Box 66
FIN-20521, Turku
Finland

Kharlamb, A
Department of Biological Sciences
Allergan Inc
Irvine, CA 92715
USA

Kobilka, B K
Division of Cardiovascular Medicine
Howard Hughes Medical Institute
Stanford University Medical School
Stanford, CA 94305
USA

Kubler, E
The Whitehead Institute for Biomedical
  Research
9 Cambridge Centre
Cambridge, MA 02142-1479
USA

Kukkonen, J P
Department of Biochemistry and
  Pharmacy
Abo Akademi University
Biocity, P O Box 66
FIN-20521, Turku
Finland

Lanier, S M
Dept. Cell & Mol. Pharm. & Exp.
  Therapeutics
Medical University of South Carolina
171 Ashley Avenue
Charleston, SC 29425-2251
USA

Lee, A
Neuroscience Graduate Program & Dept
  of Pharmacology
University of Virginia Health Sciences
  Centre
Charlottesville, VA 22908
USA

Lefkowitz, R J
Howard Hughes Medical Institute
Duke University Medical Centre
Durham, NC 27710
USA

Li, S
The Whitehead Institute for Biomedical
  Research
9 Cambridge Centre
Cambridge, MA 02142-1470
USA

Li, Y-O
Division of Pharmacological Sciences
SmithKline Beecham Pharmaceuticals
709 Swedeland Road
King of Prussia, PA 19406
USA

Liggett, S B
Departments of Medicine and
  Pharmacology
University of Cincinnati College of
  Medicine
231 Bethesda Avenue
ML 670564, Room 7511
Cincinnati, OH 45267-2564
USA

Limbird, L E
Vanderbilt University Medical Centre
Department of Pharmacology
Nashville, TN 37232-6600
USA

Lin, S
Howard Hughes Medical Institute
Stanford University Medical School
Stanford, CA 94305
USA

Lisanti, M P
The Whitehead Institute for Biomedical
  Research
9 Cambridge Center
Cambridge, MA 02142-1479
USA

Lynch, K R
Department of Pharmacology
Jordon Hall Box 448
Health Sciences Center
University of Virginia
Charlottesville, VA 22908
USA

Macdonald, T
Department of Chemistry
University of Virginia
Charlottesville, VA 22901
USA

Macmillan, L B
The Department of Pharmacology
Vanderbilt University
Nashville, TN 37232-6600
USA

Manji, H K
Schizoprenia and Mood Disorders Program
WSU School of Medicine
UHC 9B, 4201 St Antoine Boulevard
Detroit, MI 48201
USA

Manlapaz, C
Department of Biological Sciences
Allergan Inc
Irvine, CA 92715
USA

Maze, M
Department of Anesthesia
Stanford University School of Medicine
Palo Alto, CA 94305
USA

Milligan, G
Molecular Pharmacology Group
University of Glasgow
Glasgow, G12 8QQ
UK

Mizobe, T
Department of Anesthesia
Stanford University School of Medicine
Palo Alto, CA 94305
USA

Munk, S
Department of Chemical Sciences
Allergan Inc
Irvine, CA 92715
USA

Naselsky, D P
Division of Pharmacological Sciences
SmithKline Beecham Pharmaceuticals
709 Swedeland Road
King of Prussia, PA 19406
USA

Näsman, J
Department of Biochemistry and
  Pharmacy
Abo Akademi University
Biocity, P O Box 66
FIN-20521, Turku
Finland

Neubig, R R
Department of Pharmacology
University of Michigan
1301 MSRB III
Ann Arbor, MI 48109-0626
USA

Nieves, A
Department of Biological Sciences
Allergan Inc
Irvine, CA 92715
USA

Padillo, E
Department of Biological Sciences
Allergan Inc
Irvine, CA 92715
USA

Piascik, M T
Department of Pharmacology
University of Kentucky
A B Chandler Medical Centre MN-305
Lexington, KY 40536-0001
USA

Potter, W Z
Schizoprenia and Mood Disorders Program
WSU School of Medicine
UHC 9B, 4201 St Antoine Boulevard
Detroit, MI 48201
USA

Regan, J W
Department of Pharmacology and
  Toxicology
College of Pharmacy
University of Arizona
Tucson, AZ 85721
USA

Renvaktar, A
Department of Biochemistry and Pharmacy
Abo Akademi University
Biocity, P O Box 66
FIN-20521, Turku
Finland

Richman, J G
Department of Pharmacology &
  Toxicology
College of Pharmacy
University of Arizona
Tucson, AZ 85721
USA

Rosenthal, A
Department of Clinical Research
Allergan Inc
Irvine, CA 92715
USA

Ruffolo, R R
Pharmacological Sciences, UW2523
SmithKline Beecham Pharmaceuticals
709 Swedeland Road
P O Box 1539
King of Prussia, PA 19406-0939
USA

Runde, E
Department of Biological Sciences
Allergan Inc
Irvine, CA 92715
USA

Samama, P
Howard Hughes Medical Institute
Duke University Medical Centre
Durham, NC 27710
USA

Sato, M
Department of Pharmacology
Medical University of South Carolina
Charleston, SC 29425
USA

Scherer, P E
The Whitehead Institute for Biomedical
  Research
9 Cambridge Centre
Cambridge, MA 02142-1479
USA

Schmidt, M E
Schizoprenia and Mood Disorders
  Program
WSU School of Medicine
UHC 9B
4201 St Antoine Boulevard
Detroit, MI 48201
USA

Smith, M S
The Department of Pharmacology
Vascular Biology Research Group
University of Kentucky
College of Medicine
Lexington, KY 40536
USA

Soini, S
Department of Biochemistry and
  Pharmacy
Abo Akademi University
Biocity, P O Box 66
FIN-20521, Turku
Finland

Song, K S
The Whitehead Institute for Biomedical
  Research
9 Cambridge Centre
Cambridge, MA 02142-1479
USA

Spada, C
Department of Biological Sciences
Allergan Inc
2525 DuPont Drive
Irvine, CA 92715-1599
USA

Starke, K
Albert-Ludwigs-Universität-Freiburg
Hermann-Herder-Strasse 5
D-7800 Freiburg i.Br.
Federal Republic of Germany

Tang, Z
The Whitehead Institute for Biomedical
  Research
9 Cambridge Centre
Cambridge, MA 02142-1479
USA

Taylor, J M
Department of Pharmacology
University of Michigan
1301 MSRB III
Ann Arbor, MI 48109-0626
USA

Toews, M L
Department of Pharmacology
University of Nebraska Medical Center
600 South 42nd Street
Omaha, NE 68198-6260
USA

Wade, S M
Department of Pharmacology
University of Michigan
1301 MSRB III
Ann Arbor, MI 48109-0626
USA

Walt, J
Department of Clinical Research
Allergan Inc
Irvine, CA 92715
USA

Wheeler, L
Vice President
Allergan Inc
2525 DuPont Drive
Irvine, CA 92715-1599
USA

Wilson, M
Molecular Pharmacology Group
University of Glasgow
Glasgow, G12 8QQ
UK

Wozniak, M
Department of Pharmacology
Vanderbilt University School of
  Medicine
Nashville, TN 37232-6600
USA

Wu, G
Department of Pharmacology
Medical University of South Carolina
Charleston, SC 29425
USA

Yang, S-Z
Department of Pharmacology
University of Michigan
1301 MSRB III
Ann Arbor, MI 48109-0626
USA

Zhu, S-J
Department of Pharmacology
University of Nebraska Medical Center
600 South 42nd Street
Omaha, NE 68198-6260
USA

ACKNOWLEDGEMENTS

We would like to acknowledge the generous financial support provided by Orion Corporation, ORION-FARMOS, Farmos Research (Dr. Juha Matti Savola), Yamanouchi Pharmaceutical Company, LTD (Dr. Toichi Takenaka) and SmithKline Beecham (Dr. Robert Ruffolo). Funds were also donated by ALLERGAN (Dr. Larry Wheeler) ZENECA Pharmaceuticals (Dr. David U'Prichard), Bristol Meyers Squibb (Dr. Perry Molinoff), Research Biochemicals International (Dr. John L. Neumeyer) and Elsevier Trends Journals (Dr. Debra Girdlestone). Many important scientific endeavors are made possible by such institutions and dedicated individuals and we appreciate their efforts.

Of course, the meeting would not have happened (on time!) if we didn't have the energy of Kay Croker and the ASPET office, to whom we are also indebted. We thank the program committee of the American Society of Pharmacology and Experimental Therapeutics for supporting the colloquium initiative and we appreciate the effort of the staff at The Gordon and Breach Publishing Group for pushing this book to completion. We also recognize the assistance of our staff (Sybil Moore, Ester Stuart, Carolyn Pinkham, Jennifer McConnell and student/fellow volunteers) for all the detailed arrangements critical to a successful meeting. Of course, the quality of scientific exchange at such events depends not only upon the organizers but also the efforts of the participants and attendees. Such individuals made the key contribution to the colloquium and we truly appreciate their scientific excellence.

MEETING SUMMARY

Utilizing the α_2-adrenergic receptor (α_2AR) family as a template for G-protein coupled receptors in general, this meeting surveyed the receptor molecule from various perspectives, including the ligand binding pocket, receptor domains important for G-protein coupling, regulation of receptor function and receptor trafficking. Complementing this molecular context was a focus on the development of subtype-selective ligands for α_2AR receptors based on their anticipated therapeutic utility in a number of clinical settings.

Receptor Structure Critical for Catecholamine Binding and Models for Receptor Activation

The introductory session focused on the binding of ligands to the α_{2A}AR subtype, and the consequences of this binding on G-protein activation. Insights into the nature of the ligand binding pocket of the receptor are generated by various strategies, including conformational analysis of ligands. Dr. Vic Cockcroft (Orion-Farmos) exploited molecular dynamic simulations as a means to reveal multiple potential ligand conformations. By this approach, one can estimate the time that a ligand spends in any one particular conformation. Dr. Cockcroft compared the predicted residence times for structurally-similar ligands in a given conformation with the K_i and EC_{50} values determined for these ligands in radioligand binding and functional assays. Although not yet realized, one interesting possibility is that agonists of high efficacy or affinity could be predicted by determining the residence time of the ligand in particular conformational states. However, it has yet to be established whether a short or long residence time is associated with receptor activation.

Dr. Paul Heible (Smith, Kline and Beecham) related the molecular structure of the α_2AR that might account for enantiomeric differences in ligand affinity to the Eason-Stedman hypothesis. Mutation of serine 165 in the human α_{2A}AR did not alter the enantiomeric difference, whereas mutation of serine 90 and 419 resulted in selective reduction of the affinity of the receptor for the active, (–) enantiomer. These findings extend the current understanding of the catecholamine binding pocket of the α_2AR and the amino acids proposed to interact with catecholamine ligands.

One key extension of findings on the catecholamine binding pocket is the delineation of conformational changes in the receptor protein subsequent to agonist binding. Dr. Üllrich Gether (colleague in the laboratory of Dr. Brian Kobilka, Stanford University) utilized fluorescence spectroscopic analysis of the β_2-adrenergic receptor, another catecholamine-binding receptor, purified from SF9 cells to reveal conformational changes associated with ligand binding. The purified, functional receptor was labeled with the fluorescent cysteine-modifying agent, IANBP and the change in fluorescence elicited by addition of receptor ligands was related to their respective ability to activate adenylyl cyclase. Observed changes in fluorescence elicited by agonist could be due to vertical or horizontal movements of the transmembrane domains of the receptor in relation to the inside versus the outside of the cell or, alternatively, rotational events of the transmembrane spans within the bilayer. Conformational changes induced by agonist occupancy of the β_2AR were evaluated by examining changes in fluorescence elicited by receptor structures created by sequential mutation of each of the cysteines in the β_2AR. The data suggest that cysteine 125 in transmembrane (TM) 5 and cysteine 283 in TM 6 are critical for the agonist-induced fluorescent changes and that the binding of agonists initiates rotational events involving these two transmembrane spans. Dr. Gether extrapolated these findings into a multi-state model for receptor conformations, where the receptor exists in four global states: fully active, fully inactive, partially active, partially inactive. This and other comments during the conference emphasized that G-protein coupled receptors are dynamic entities that likely oscillate between numerous conformations, only some of which are currently detectable using existing experimental strategies and only some of which are capable of activating G-proteins.

Multiple conformational states of G-protein coupled receptors were discussed in some detail by Dr. Philip Samama (Duke University) in the context of the extended ternary complex model. This model accommodates a variety of emerging findings from a number of laboratories, including that (1) agonist-independent activity is observed in some cell systems expressing higher receptor density and (2) the ability of receptor antagonists to behave either as null antagonists or as negative antagonists (i.e., inverse agonists). The extended ternary coupled model evolved from the serendipitous discovery that 'gain of function' mutations of β_2AR could be observed *in vitro* by introducing sequences from the $\alpha_{1B}AR$ into the carboxy terminal end of the third intracellular loop of the β_2AR. These mutations led to a constitutively activated (i.e. agonist-independent) β_2AR whose activity was suppressed by certain antagonists. An intriguing property of these constitutively active receptors is that they have a higher affinity for agonists than the wildtype receptor, whereas the receptor affinity for antagonists is not modified. Importantly, this increase in receptor affinity for agonists in constitutively active β_2AR does not require the involvement of G-proteins, as it is detected following detergent extraction, which physically dissociates the receptor from the G-protein. These findings required an extension of the original ternary complex model to include an allosteric, or conformationally-altered, state representing interconversions of the receptor between basal (R) and inactivated (R*) states. Dr. Samama proposed that it is perhaps best to consider the role of an unoccupied receptor as a 'brake' on signal transduction and the role of agonist as a means to eliminate this 'brake' on receptor activity. In many ways, this allosteric model resembles features of the model presented by Üllrich Gether, discussed above, except that the number of conformational states in the extended ternary complex model is defined, whereas the number of potential conformational states in the allosteric model proposed by Ü. Gether is virtually infinite.

One important element of the extended ternary complex model is the provision of formal descriptors for quantifying the extent of agonist-independent activation of a receptor in different cellular settings, particularly receptors rendered constitutively active by mutations occurring naturally and leading to human disease. As the mutations leading to these defects in human disease (e.g., retinitis pigmentosa, precocious puberty, thyrotoxicosis, and others) occur throughout the seven transmembrane-spanning structures and not just in the carboxy-terminal sequences of the third intracellular loop, the structural basis for the ability of these receptors to achieve a functional R^* conformation even in the absence of agonist has yet to be revealed.

Structural Regions of the α_2AR Involved in G-protein Coupling

The importance of the carboxy terminal region of the third intracellular loop of the α_2AR in activation of G-proteins was also affirmed by findings obtained using complementary biochemical strategies, involving receptor structure-based peptides (Dr. Richard Neubig and his colleagues, University of Michigan). One particular peptide (dubbed the Q peptide), corresponding to amino acids 361-373 in the carboxy-terminal end of the third cytoplasmic loop of the α_2AR, effectively inhibited receptor-mediated as well as mastoparan activation of G protein. Interestingly, peptide Q appears to be capable of interacting with both the α and β subunits of G-proteins. Thus, derivatized peptide Q photoincorporated not only into a 2 KD fragment from the amino terminus of $G_{i\alpha}$, consistent with earlier modeling and recent structural data for the interface between receptors and heterotrimeric G-proteins, but also into the β subunit of G-proteins. These findings provide direct evidence for the expectation that several physical contact points exist between receptors and heterotrimeric G-proteins and thus could contribute to the detection of subtle differences in ligand structure of agonists, partial agonists, antagonists, and inverse agonists, translating these into variable transfer of signal from receptor to G-proteins.

The carboxy-terminus of the third intracellular loop of $\alpha_{2A}AR$ was also implicated in the selective coupling of this receptor subtype to G_i *versus* G_s (Eason and Liggett, University of Cincinnati). Deletion of amino acids 221 to 231 of the $\alpha_{2A}AR$ led to a loss of G_s but not G_i coupling. Similar results were obtained when this amino acid segment in the $\alpha_{2A}AR$ was substituted with a similar region from the β_2AR or 5HT-1A receptor. To date, activation of G_s by α_2AR by a pertussis-toxin insensitive pathway has been reported primarily in heterologous systems expressing recombinant receptors. It is not known if direct coupling to G_s occurs *in vivo*, as activation of adenylyl cyclase by α_2AR could also occur *via* liberation by $\beta\gamma$ subunits of G_i and G_o or, alternatively, *via* alteration of Ca^{++} availability and/or the phosphorylation state of the isoform(s) of the enzyme present in the target cell of interest. Now that the molecular identification of adenylyl cyclase isoforms is possible in a cell-specific manner, these studies can be undertaken.

α_2AR and associated signaling complexes

The studies reported by Dr. Lanier (University of South Carolina) described efforts to reveal the mechanisms and perhaps novel proteins that regulate signal transfer from α_2AR to G-proteins. These studies demonstrated that α_2AR co-expressed to comparable densities in PC12 (cultured rat pheochromatcytoma) cells versus NIH-3T3 cells exhibit unexpected reciprocal relationships between high affinity agonist binding (optimal in 3T3-L1 cells) and receptor-stimulated GTPase activity, which occurs to a much greater 'fold' stimulation in PC12 cells than in NIH-3T3 cells. These differences do not appear to be readily accounted for by differences in G-protein density or by differing G-protein composition in the two heterologous cell 'backgrounds', suggesting that they instead reflect the existence of an 'amplifier' of α_2AR mediated signal transduction that is uniquely enriched in PC12 cells. This signal-transducing enhancement activity is also observed in NG 108-15 neuroblastoma glioma cells, another cell line including genetic elements of neuronal origin. This activity was partially purified from membrane extracts by DEAE-ion exchange chromatography. In complementary studies to identify molecules that might facilitate receptor-G protein-coupling, a GST-fusion protein strategy was exploited to identify cytosolic proteins that interact with the predicted third intracellular loop of the α_{2A}AR. Although early in their evolution, these studies suggest the possibility that novel molecular players will be identified that influence functionally productive interactions among receptor-G protein-effector complexes.

Dr. Michael Lisanti (MIT) presented data suggesting that a higher order of specificity of signal transduction might be achieved by enrichment of G protein signalling complexes in molecular organelles such as caveoli. Caveoli are flask-shaped cell surface 50–100 nm invaginations which were first proposed to play roles in endocytosis and pinocytosis and enriched in a protein caveolin (Mr ~ 28,000), initially identified as a major C-src substrate in RSV-transformed cells. Purification of these caveolin-rich structures has revealed the association of caveolin with a number of signal transducing molecules, including C-src, C-yes, lyn, jyn, lck, C-fgr and JAK2, all non-receptor tyrosine kinases. In addition, several types of G protein α and β subunits have been identified in different preparations of caveolin-rich structures. Other reports in the literature have demonstrated the association of β-AR and muscarinic acetylcholine receptors with caveolin following ligand-induced receptor redistribution. This cooperative association of signal transducing molecules with a cell surface 'organelle' structure that participates in endocytotic events has led to the development of the 'caveoli signalling hypothesis'. This hypothesis proposes that ligand occupancy of certain receptor populations results in their enrichment in caveoli, where they become concentrated with transducers of signal transduction pathways. This hypothesis leads to certain testable predictions, including the prediction that the transducing molecules that are pre-enriched in caveoli exist in an inactive state but, upon encounters with agonist-occupied receptors in caveoli, become activated and disassociated from caveolin, an expectation confirmed by data shown for $G_{s\alpha}$ by Lisanti.

The possible importance of caveoli in coordinating multicomponent signalling pathways encouraged Dr. Lisanti to identify what structural features of caveolin lead to its organization into caveolar structures. Caveolin, which possesses a hairpin-like structure in the bilayer, oligomerizes to form the caveolar structure. Molecular cloning has revealed a family of caveolin molecules (caveolin 1 and its splice variants; caveolin 2, 3; and possibly caveolin 4). Caveolin 1 and 2 have the same tissue distribution and are both co-localized in the same cells. Recent data show that varying caveolin isoforms alter GTP binding, GDP release and GTPase activation. PC12 cells do not express caveolin 1, but expression cloning strategies have suggested that a novel caveolin molecule may exist in these cells. An exciting possibility is that enrichment of receptors and G-proteins could be achieved by caveolins with subtle differences in molecular structure, and that the coordination of signal activation and inactivation might occur in such cellular subcompartments.

α_2AR coupling to diverse signaling pathways

The coupling of $\alpha_{2A}AR$ to signaling pathways other than inhibition of adenylyl cyclase was explored in several presentations. Although $\alpha_{2A}AR$ are coupled to suppression of voltage-gated Ca^{++} currents via G_o-mediated signaling in some cells, in other target cells α_2AR activation *increases* intracellular Ca^{++} concentrations. This elevation in intracellular Ca^{++} can occur via a variety of mechanisms. As described by Dr. Ackerman (Åbo Akademi University, Finland), in some but not all settings the release of intracellular Ca^{++} is pertussis toxin-insensitive, parallels inositol-PO_4 production, and can be blocked by inhibitors of phospholipase C. One target cell that manifests this profile of Ca^{++} elevation in response to α_2AR agonists is human erythroleukemia (HEL) cells, and Dr. Ackerman observed that α_2AR can accelerate inositol 1,4,5 trisphosphate production in HEL cells, albeit not to the magnitude induced by other receptor systems. An attractive model system for pursuing the molecular basis for pertussis toxin-insensitive Ca^{++} mobilization evoked by α_2AR is the SF nine (9) cell, a cultured insect cell line that apparently has all of the relevant GTP-binding proteins to mediate α_2-AR activation of Ca^{++} mobilization. Furthermore, antibodies developed against rat G-proteins also recognize G-proteins expressed in SF9 cells. Ackerman observed that dexmedetomidine, norepinephrine, clonidine and UK14,304 all activate Ca^{++} mobilization in these cells, whereas L-metetomidine and rauwolsine act as negative antagonists to suppress basal (agonist-independent) increases in Ca^{++} accumulation.

When mobilization of Ca^{++} by α_2AR is examined in Chinese hamster ovary (CHO) cells, there is a subtype selectivity in Ca^{++} mobilization, such that $\alpha_{2C}AR$ elicit only small activation of Ca^{++} signals whereas $\alpha_{2B}AR$ elicit strong Ca^{++} mobilization signals. (The $\alpha_{2A}AR$ has not yet been studied in this context in CHO cells.) The $\alpha_{2B}AR$-evoked Ca^{++} mobilization in CHO cells appears to be pertussis-toxin-sensitive, suggesting that phospholipase C-mediated inositol 1,4,5 trisphophate production in CHO cells relies on $\beta\gamma$ subunits liberated upon activation of G_i and/or G_o proteins.

Dr. Graeme Milligan discussed the intersection of α_2AR-activated G-protein coupled pathways with the activation of p^{21} ras and the tyrosine kinase cascade. In Rat 1 cells, Milligan and colleagues (Glasgow, Scotland) demonstrated that α_2AR activate both G_{i2} and G_o G-proteins, using an ingenious strategy to identify receptor-activated G proteins, namely the ability of agonist-stimulated loading of G_i and G_o proteins with GTP to transform G_i and G_o into cholera toxin substrates. Quantitation of cholera toxin-catalyzed ^{32}P-ADP-ribosylation of GTP-loaded G_i and G_o variants allows identification of those G-proteins activated by α_2AR, after G-protein isolation by immuno-precipitation; to better detect the agonist-stimulated changes in the ^{32}P-ADP rebosylation signal. Confirming earlier findings by Moolenaar and colleagues, Milligan described findings that α_2AR introduction into Rat 1 cells was paralleled by an ability to detect agonist-stimulated α_2AR loading of p^{21} ras with GTP. Activation of the ras pathway is pertussis toxin-sensitive, implicating G, or G_o in the pathway, although ras activation does not involve suppression of cyclic AMP synthesis. The downstream consequences of ras activation, including phosphorylation of map kinase and activation of the map kinase pathways, are also observed. Inhibition of PI3 kinase activity completely suppresses activation of the map kinase cascade by lysophophatidic acid (another receptor-mediated, pertussis toxin-sensitive pathway) and by α_2AR. In contrast, the effects of PDGF on ras and map kinase signaling in Rat 1 cells are not pertussis toxin-sensitive. The implications of these findings for α_2AR stimulation of proliferation and mitogenesis in target cells *in vivo* remains to be established.

Regulation of receptor – G protein coupling by desensitization mechanisms

Dr. Jeffrey Benovic (Thomas Jefferson University) discussed the regulation of G-protein coupled receptors by agonist-evoked phosphorylation. Agonist-induced receptor phosphorylation is associated with uncoupling of the receptor from G-protein activation and can result from second messenger activation (heterologous desensitization pathways) or from agonist occupancy of the receptor (homologous desensitization mechanisms). In the latter case, agonist occupancy of the receptor leads to a conformational change in the receptor and parallel activation of G-proteins that, upon liberation of $\beta\gamma$ subunits, fosters the association of a G protein-coupled receptor-directed kinase (GRK) with the membrane and phosphorylation of the agonist-occupied receptor. Functional uncoupling of receptors and G-proteins parallels this phosphorylation event. Further hyper-phosphorylation of agonist-occupied G protein-coupled receptors is noted following association of a molecule, dubbed arrestin, with the GRK-phosphorylated receptor. This paradigm was first described for rhodopsin, rhodopsin kinase and visual arrestin; more recently it has been studied in detail for β_2AR in the laboratories of Benovic (Thomas Jefferson University) and of Lefkowitz (Duke University).

Molecular cloning studies have identified a large family of GRKs: GRK1 is equivalent to rhodopsin kinase, GRK_2 is equivalent to βARK1 (β-adrenergic receptor kinase 1). The isoforms of GRKs are highly homologous in their amino terminal and catalytic domains, but differ in their C-termini. GRK2 and 3 have a pleckstrin homology domain in their C-terminus that appears to be important for GRK association with $\beta\gamma$ subunits of G proteins and for stabilizing GRK association with the plasma membrane. In contrast, GRK1 is farnesylated and GRK4 and 6 are palmitoylated. This post-translational acylation may subserve the same function of stabilizing GRK association in the bilayer with agonist-occupied G-protein-coupled receptors. A large number of receptors are phosphorylated upon agonist occupancy by GRKs including $\alpha_{2A}AR$, $\alpha_{2C}AR$, $M_2Ach R$, and A1 adenosine receptors coupled to G_i; β_1- and β_2-AR, D1 dopamine receptors, A2 adenosine receptors, and LH\hCG receptors coupled to G_s; CCK, GRP, substance P, thrombin, and M_3AchR, $\alpha_{1B}AR$, FMLP, C5A, and odorant receptors coupled to $G_{\alpha q/11}$ and to phospholipase C. What remains to be established is the specificity of each of these GRKs and their molecular mechanisms for interaction with particular receptors coupled to differing G protein-coupled signalling pathways.

Dr. David Bylund (University of Nebraska) presented a number of criteria that are consistent with the interpretation that α_2AR subtypes may give rise to cell-specific regulation. The functional relevance of α_2AR receptor subtypes has always been queried, particularly since the potency of the *in vivo* agonists, epinephrine and norepinephrine, is virtually indistinguishable at all three α_2AR subtypes. One possible explanation for the existence of receptor subtypes, posed by Dr. Bylund and other investigators, is that differential regulation of these subtypes might exist. *In vitro* correlates of this regulation are differential sensitivity to buffer components and ions, but the *in vivo* relevance of the existence of α_2AR subtypes may be differences in desensitization and subsequent 'down regulation' of each α_2AR subtype. Agonist-induced receptor down regulation (i.e., decrease in functional receptor density) of α_{2B}AR is observed at low agonist occupancy and to a greater extent than for α_{2A}AR; down regulation of the α_{2A}AR is detected only after 24 hours and at high agonist concentrations. The α_{2C}AR, studied in the endogenously-expressing opossum kidney (OK) cell line, and α_{2B}AR heterologously expressed in NIH 3T3 cells show similar agonist concentration-agonist response curves for receptor down regulation, albeit the α_{2C}AR is slightly more sensitive than the α_{2B}AR in the loss of receptor density. Importantly, the EC_{50} for norepinephrine in decreasing α_{2B}- or α_{2C}-AR density (10 and 90 nM respectively), is strikingly lower than the EC_{50} observed for the α_{2A}AR (30 μM) to evoke down-regulation. An interesting interpretation of the findings from Dr. Bylund's laboratory is that the conformation of the receptor, important for signaling, may be different from that critical for down-regulation, based on the relative potency of agonists in effecting down regulation *versus* receptor activation. Additional data from Dr. Bylund's laboratory are consistent with the hypothesis that the molecular mechanism of agonist-induced down-regulation is an acceleration of receptor degradation rather than an attenuation of α_2AR synthesis; Dr. Bylund interpreted his findings to suggest that changes in α_2AR degradation rate don't necessarily produce down regulation but do assist in maintaining a decreased receptor density.

Dr. Steve Liggett (University of Cincinnati), examining α_2AR desensitization pathways, demonstrated that occupancy of all three α_2AR subtypes is quickly followed by agonist-promoted receptor phosphorylation, presumably via the GRKs described by Dr. Benovic. The sequences on the α_{2A}AR phosphorylated by GRKs appeared to be the polyserine residues in the third intracellular loop (imbedded in the LEESSSS sequence, ser 296–299). Findings from the intact cell phosphorylation assay utilized in their laboratory as a measure of *in situ* receptor regulation demonstrate that receptor-dependent phosphorylation requires a high concentration of agonist, is inhibited by heparin (an inhibitor of GRKs), is unaffected by concanavalin A (a tool utilized to prevent α_2AR endocytosis), and is paralleled by a loss of functional coupling of α_2AR to G proteins. Mutant receptors that fail to phosphorylate also fail to desensitize, although they are like wildtype α_2AR in their ability to inhibit adenylyl cyclase. Phosphorylation of all 4 serines on the α_{2A}AR appears critical for functional desensitization, since replacement of individual Ser residues results in an incremental loss of phosphorylation, but the replacement by mutagenesis of any of these serines with alanine results in abolition of agonist-promoted desensitization. Recent studies have demonstrated that the α_{2C}AR also is phosphorylated following agonist occupancy, and the sites for phosphorylation have been identified and recently reported. What remains to be established is which GRK phosphorylates each of the α_2AR . *In vitro*, GRK 2 and 3 (also known as βARK 1 and 2) are able to phosphorylate the α_2AR subtypes, whereas GRK 5 and 6 do not; perhaps this can be attributed to the pleckstrin homology domain that exists in the C-terminus of GRK 2 and 3, but is absent in the C-terminus of GRK 5 and 6, as discussed earlier in the summary of findings presented by J. Benovic.

Localization of α_2AR in target cells

The trafficking pathways utilized by α_2AR subtypes to achieve polarized expression in epithelial cells, using Madin-Darby canine kidney cells as a cultured cell model, were reported by Magda Wozniak (colleague in the laboratory of L.E. Limbird, Vanderbilt).

All three α_2AR subtypes are expressed on the lateral subdomain of the basolateral surface at steady state, as revealed by confocal microscopy and surface biotinylation strategies that quantitate the distribution of photoaffinity-labeled α_2AR on the apical *versus* basolateral surface. However, pulse-chase studies of metabolically-labeled MDCK cells demonstrate that the α_2AR subtypes are targeted to this lateral surface *via* different mechanisms. Whereas α_{2A}AR and α_{2C}AR are directly delivered to lateral membrane, the α_{2B}AR are delivered randomly to the apical and basolateral surfaces, but are enriched at steady state on the basolateral surface *via* selective retention in that membrane. Thus, α_{2B}AR are rapidly ($t_{1/2}$ = 15–30 min) removed from the apical surface, in contrast to the longer retention of the α_{2B}AR on the basolateral surface, when the surface half-life is 10–12 hours. It is not yet known if the α_{2B}AR removed from the apical surface is degraded or is redistributed to the basolateral surface. In addition to expression on the basolateral surface, the α_{2C}AR in MDCK cells also demonstrates significant intracellular localization, corroborating previous reports by von Zostrow and Kobilka for α_{2C}AR expression in COS and HEK293 transformed renal epithelial cell lines.

Data presented by David Daunt (colleague in the laboratory of B. Kobilka, Stanford) further emphasized the unique cellular distribution properties of the α_{2C}AR subtype. Immunocytochemical detection of α_{2C}AR localization in transfected human embryonal kidney (HEK293) cells confirmed that the α_{2C}AR is found on the surface as well as in intracellular vesicles. By examining the localization of α_{2A}/α_{2C}AR chimeras, the intracellular distribution of the α_{2C}AR was revealed to be due to a stretch of hydrophobic amino acids in the amino-terminus of the α_{2C}AR.

Dr. Daunt also reported on studies designed to examine agonist-evoked receptor distribution of α_2AR subtypes. While α_{2B}AR undergo rapid sequestration as a consequence of cell exposure to epinephrine, the α_{2A}AR does not. Examination of agonist-evoked redistribution of α_{2A}/α_{2B}AR chimeras suggest that sequestration is determined by amino acids in the carboxy tail of the α_{2B}AR sequence. A single mutation of histidine into alanine or glutamine in α_{2A}AR carboxy-tail creates a receptor that rapidly internalizes in response to agonist, raising the possibility that the histidine acts as a brake for sequestration in the native α_{2A}AR. However, mutation of the corresponding glutamine into histidine in the α_{2B}AR sequence does *not* alter internalization, suggesting that other domains are also involved in the sequestration process.

J. Richman (colleague from the laboratory of J. Regan, University of Arizona) described findings utilizing antibodies against the native α_{2A}AR subtypes to reveal the localization of endogenous α_2AR in various target cells. Labeling of cultured rat spinal cord neurons with antibodies raised against the predicted third intracellular loop of the three α_2AR subtypes demonstrated that α_{2A} and α_{2B}AR subtypes were co-expressed in a single neuron. Both α_{2A}- and α_{2B}-AR were expressed in the cell body. Localization of α_{2A}AR in the somadodendritic membrane of neurites α_{2A}AR localization is punctate, whereas the α_{2B}AR expression appears segmented. A comparative study in their laboratory of different cell types (COS cells, fetal cardiomyocytes) indicated that the subcellular localization of α_2AR subtypes and their sequestration in response to agonist exposure varies considerably with the different target cell evaluated.

Dr. Kevin Lynch (University of Virginia) described the subcellular localization of the α_{2A}AR subtype in brain slices, using an affinity-purified rabbit polyclonal antibody directed against a polypeptide corresponding to 47 amino acids of the predicted third intracellular loop of the α_{2A}AR. Examination of immunoreactive neurons revealed diffuse staining in cell bodies as well as punctiform intracellular staining. The complementary examination of GT1 cells (an immortalized GnRH-secreting hypothalamic cell-line expressing α_2AR) confirmed the observations made *in situ*. Studies at the electron microscopic level demonstrated that the intracellular clusters of α_{2A}AR immunoreactivity do not correspond to endosomes or lysozomes, but parallel with the profile detected for immunolocalization of the motor protein kinesin.

In vivo evaluation of α_2AR function

Dr. Michael Piascik reported on the cardiovascular effects of α_2AR and α_1AR *in vivo*. In collaborative studies with Leigh MacMillan (colleague of Lee Limbird, Vanderbilt University), the central hypotensive effects of clonidine and its analog, UK14,304, were observed to be disrupted in a D79N homozygous α_{2A}AR mutant mouse model developed by Leigh MacMillan. These findings are consistent with the interpretation that imidazoline structural agents, like UK14,304, mediate most of their central effects on blood pressure *via* α_{2A}AR, without necessarily requiring the involvement of distinct imidazoline receptors. Peripheral vascular effects of α_2 agonists on blood pressure, however, appear to be due principally to receptors of the α_{2B}AR subtype, since vascular contractility is still detected in aortic rings derived from wildtype and D79N α_{2A}AR mutant animals and is sensitive to prazosin (which blocks α_{2B}AR and α_{2A}AR in addition to α_1AR), and since disruption of expression of the α_{2B}AR subtype using gene targeting strategies (Richard Link, Kobilka laboratory) eliminated the transient vasoconstriction evoked by intravenous administration of α_2AR agonists. Taken together, the findings to date in genetically-modified mice suggest that instantaneous changes in peripheral vasoconstriction evoked *in vivo* may be mediated principally, at least in the mouse, by the α_{2B}AR subtype, with contributions by the α_{2A}AR subtype, whereas centrally-regulated hypotension is mediated via the α_{2A}AR subtype.

Dr. Piascik described novel technology he has developed for successful disruption of expression of α_1AR focusing specifically on the α_{1D}AR subtype using antisense strategies. Antisense oligonucleotides corresponding to the translation initiation start site of the α_{1D}AR were introduced into a thermal-sensitive gel that, upon warming to 37%, solidifies *in vivo*. Introduction of α_{1D}AR-directed oligonucleotide led to loss of α_1AR-mediated vasoconstriction, implicating a role for the α_{1D}AR subtype in vasoconstriction of the femoral artery. These strategies suggest the possibility of *in vivo* evaluation of the role of varying receptor subtypes in different vascular beds using this technique for regional introduction of antisense oligonucleotides.

Current and anticipated therapeutic utility of a_2AR agents

The conclusion of the program focused on therapeutic implications of α_2AR in depression (Dr. Manji, Wayne State University), glaucoma (Larry Wheeler, Allergan) and analgesia/anesthesia (Maze, Stanford). Bipolar depression, by definition, includes elements of both mania and depression. Most antidepressant agents administered clinically in the past have been targeted toward presynaptic elements that block norepinephrine reuptake and, more recently, serotonin reuptake and thus provide greater extent and duration of neurotransmitter availability. However, problems with these therapies include their major side effect profiles. These include sedation and orthostatic hypotension, due to interaction of norepinephrine with postsynaptic α_1 and α_2 adrenergic receptors, and constipation, dry mouth, and urinary retention due to interaction of norepinephrine, when available in excess, at H1 histamine receptors. A very dangerous side effect of currently available blockers of neurotransmitter transporters is their life-threatening consequences following overdose, a particular problem for patients who, when depressed, are suicidal in nature. Another limitation to currently available antidepressant therapy is the long delay until action, sometimes taking 7–10 days or as long as several weeks to effect major antidepressant actions. Furthermore, 20–30% of patients never respond to any of the currently available antidepressant agents. Since plasma norepinephrine levels decline in depression, readily revealed in the cold pressor test of depressed patients, one hypothesis is that antidepressant therapy might be achieved using α_2AR antagonists; the availability of α_2AR antagonists at neuronal terminals would increase firing, and thereby lead to more norepinephrine availability. Clinical studies with intravenous idazoxan, an α_2AR antagonist, reveal an elevation of norepinephrine in the plasma of almost every patient examined, although there is considerable inter-individual variability. Interestingly, further evaluation of this variability revealed gender-specific differences in response to idazoxan. These differences could be minimized by adjunct treatment of women with GnRH long-acting agonists, which suppress endogenous estrogen production. This adjunct therapy led to an improvement in antidepressant response to idazoxan in female subjects. Deoxyglucose measurements, used as an index of neuronal activity, indicated that idazoxan infusion caused significant improvement in attention in patients with bipolar or unipolar depression and in color-word recognition, a measure of cognition. Treatment at the National Institutes of Health of 20 patients for whom nothing had worked therapeutically in the past showed an anomalous distribution: seven of the ten patients who had bipolar depression manifested increased nonadrenergic activity, whereas only two of the nine unipolar patients did. These data, albeit preliminary, suggest that α_2AR antagonists might serve as a clinically effective paradigm for the treatment of depression, particularly in individuals who suffer from bipolar depression. Since idazoxan also binds to imidazoline I_2-binding sites in addition to α_2AR, the contribution of these receptors to the beneficial effects of idazoxan cannot be excluded. Experiments currently being performed with ethoxy-idazoxan, which is devoid of I_2-site interaction, should be particularly informative for refining further development of strategies for the treatment of depression.

Another fertile area for α_2AR drug development is in analgesia and anesthesia, discussed by Mervin Maze (Stanford University). α_2AR agonists possess anesthetic sparing, analgesic and anxiolytic properties. Genetic strategies that have created null mutations for the α_{2B}AR and α_{2C}AR subtypes have shown little impact on anesthesia and sedation (Maze in collaboration with Kobilka and colleagues); however, the D79N α_{2A}AR homozygous mice (Leigh MacMillan, laboratory of Lee Limbird at Vanderbilt University) no longer respond to the anesthetic sparing effects of dexmedetomidine, do not demonstrate a suppressed righting reflex characteristic of α_2AR agonist-induced sedation, and have altered analgesic response to α_2AR agonists when evaluated in a ramped hot plate assay. Taken together, these findings suggest that the effects of α_2AR agonists on analgesia, sedation, and anesthetic sparing actions are mediated by the $\alpha_{2A}AR$ subtype. Other studies reported by Dr. Maze are consistent with a role for suppression of Ca^{++} currents in mediating these effects of $\alpha_{2A}AR$ on sedation. *In vivo* studies performed in Dr. Maze's laboratory demonstrate considerable differences in receptor reserve for varying $\alpha_{2A}AR$ responses that may be of therapeutic importance. Findings reported also suggest that $\alpha_{2A}AR$ agonists might provide important therapy, or adjunct therapy, in alleviating chronic neuropathic pain, as revealed both in animal studies and in preclinical studies of transdermal clonidine administration to suppress hyperalgesia.

Several adrenergic ligands exert variable degrees of effectiveness in the treatment of glaucoma. The progression of glaucoma often is accompanied by increased intraocular pressure and damage to the optic disc. Since intraocular pressure creates the appropriate geometry for vision, effective pharmacological maintenance of intraocular pressure could be a clinically effective tool for treatment of glaucoma. The therapeutic goal in glaucoma is to decrease intraocular pressure to a value of 18 mm of Hg or less. Most current therapy focuses on timolol, a βAR agonist. However, the $\alpha_2 AR$ agonist brimonidine (UK14,304; 0.2%) shows promise as a potential therapeutic agent for the reduction of intraocular pressure. Studies using measurements of intraocular pressure in the rabbit as a model have compared the properties of UK14,304 to those of clonidine and para-aminoclonidine in lowering intraocular pressure without causing untoward side effects. Interestingly, UK14,304 can change fluid outflow as well as intraocular pressure, whereas PIC does not affect changes in fluid outflow, suggesting that UK14,304 potentially has two means by which it could effect successful therapeutic intervention in glaucoma. Furthermore, UK14,304 showed reduced side effects compared to the other two $\alpha_2 AR$ agonists. For example, mydriasis, one side effect of $\alpha_2 AR$ agonist treatment, is profound with PIC, noted with clonidine, but not observed for UK14,304. A perceived limitation of $\alpha_2 AR$ agonists for glaucoma therapy is the anticipated vasoconstriction, which would compromise blood flow to the optic nerve and counteract the therapeutic effects derived from reduction of intraocular pressure effected by α_2 agonists. However, studies in human retinal xenografts indicate that UK14,304 does not affect vasoconstriction in this setting, whereas PIC and clonidine do. Taken together, these studies provide promising structural insights for the development of $\alpha_2 AR$ agonists that might evoke therapeutically-effective decreases in intraocular pressure with limited or absent untoward side effects.

CONCLUDING PERSPECTIVES

The meeting was closed by a very provocative summary presentation by Dr. Paul Hartig (Dupont Merck). In trying to link the molecular, *in vitro*, *in vivo*, and therapeutic studies, Dr. Hartig focused on current therapeutic opportunities for $\alpha_2 AR$ agonists and future goals for biomedical research and training in this area. Dr. Hartig proposed that it was his opinion that $\alpha_2 AR$ directed ligands would have their greatest therapeutic potential in the central control of cardiovascular function, in effecting sedation and analgesia, and in protecting against cerebral ischemia. On a speculative and philosophical note, Dr. Hartig queried about the value of developing or screening for receptor subtype-targeted agents. Since subtle changes in receptor structure exist between species, and these subtle structural differences occasionally significantly modify receptor affinity for particular drug analogs, premature and hence wasteful drug screening efforts may be undertaken if prior studies do not first confirm that similar structure-activity relationships exist at the same receptor subtypes in human beings. Dr. Hartig provoked the audience to think about genetic diversity. He reminded the audience that there is approximately 99% nucleotide identity in the coding and noncoding regions of the human genome between individuals, 98–99% identity between human beings and primates, and approximately 90% identity between sequences for receptors in human beings and in rodents, a common animal model for drug screening to evaluate modulation of physiological and behavioral functions. To both be informed by and respond to emerging insights from The Human Genome Project, attendees were urged to obtain and sustain contemporary knowledge concerning molecular structure and *in vitro* cell signalling studies while similarly embracing an appreciation for advances in the understanding of the receptor subtypes that contribute to *in vivo* responses.

The conference was successful in linking molecular, cellular, *in vivo* and therapeutic insights, and also affirmed the role the α_2AR have played as a structural model to understand G protein-coupled receptor folding, trafficking, steady state localization, coupling to disparate effectors via differing G-proteins (and α *versus* $\beta\gamma$ subunits of these heterotrimeric proteins) desensitization, receptor turnover, and contributions to complex *in vivo* regulating pathways. It is anticipated that the breadth of experimentation described at the conference will be the basis not only for the design of α_2AR subtype selective drugs but also for the discovery of novel drug targets in the trafficking and signalling pathways utilized by or regulated by these receptors.

Stephen M. Lanier
Max Lafontan
Lee E. Limbird
Hervé Paris

IDENTIFICATION, CHARACTERIZATION AND SUBCLASSIFICATION OF α_2-ADRENOCEPTORS: AN OVERVIEW

J. PAUL HIEBLE,[*] ROBERT R. RUFFOLO, JR.[*] and KLAUS STARKE[†]

[*]SmithKline Beecham Pharmaceuticals, 709 Swedeland Road, P.O. Box 1539, King of Prussia, PA, USA 19406-0939, [†]Albert-Ludwigs-Universität-Freiburg, Hermann-Herder-Strasse 5, D-7800 Freiburg i.Br., Federal Republic of Germany

IDENTIFICATION OF THE α_2-ADRENOCEPTORS

Functional studies

The initial identification of the α_2-adrenoceptor was coupled to the discovery of the prejunctional autoreceptor on adrenergic nerve terminals which participates in the negative feedback system controlling norepinephrine release in response to sympathetic nerve activation (for details of the history see Starke, 1977). It had long been known that low concentrations of some α-adrenoceptor antagonists could potentiate the functional responses to sympathetic nerve stimulation (Cannon and Bacq, 1931; Jang, 1940), and that α-adrenoceptor antagonists would enhance nerve-evoked norepinephrine release (Brown and Gillespie, 1957). This phenomenon was eventually explained by the presence of presynaptic α-adrenoceptors, later called autoreceptors (Farnebo and Hamberger, 1971; Kirpekar and Puig, 1971; Langer et al., 1971; Starke et al., 1971a). Confirmation of this hypothesis was provided by the observation that exogenous norepinephrine can inhibit the nerve-evoked release of endogenous norepinephrine (Starke, 1972b). Inhibition of stimulation-evoked norepinephrine overflow was also demonstrated with the imidazoline α-adrenoceptor agonists, clonidine, naphazoline and oxymetazoline (Werner et al., 1970; Starke, 1971, 1972a) and with the thiazine α-adrenoceptor agonist, xylazine (Heise and Kroneberg, 1970).

It soon became apparent that the prejunctional α-adrenoceptor had pharmacological characteristics that were distinct from the previously evaluated postjunctional α-adrenoceptors, based on the relative potencies of a series of α-adrenoceptor agonists for inhibition of neurotransmitter release and stimulation of contractility in rabbit heart (Starke, 1972a), and for producing contraction or inhibition of neurotransmitter release in the rabbit pulmonary artery (Starke et al., 1974, 1975b). In the pulmonary artery, clonidine and α-

methylnorepinephrine were shown to act preferentially to inhibit norepinephrine release, whereas methoxamine and phenylephrine were more effective in producing vasoconstriction. Similar results were obtained in the rabbit ear artery, where clonidine and α-methylnorepinephrine were demonstrated to inhibit the vasoconstrictor response to field stimulation, a response reflecting norepinephrine release from the adrenergic nerve terminal, at concentrations lower than those required to induce vasoconstriction through direct activation of α-adrenoceptors located postjunctionally on the smooth muscle. In contrast, methoxamine and phenylephrine produced only vasoconstriction and had no effect on neurotransmitter release (Steinsland and Nelson, 1975). In both the rabbit pulmonary artery and rabbit ear artery, the endogenous catecholamines, norepinephrine and epinephrine, appeared to produce essentially equivalent responses at pre- and postjunctional α-adrenoceptors.

Phenoxybenzamine was shown to block preferentially the postjunctional α-adrenoceptor in the cat spleen and rabbit pulmonary artery (Dubocovich and Langer, 1974), although the irreversible blockade of α-adrenoceptors produced by this drug makes accurate potency comparisons in different systems difficult. Yohimbine was subsequently shown to have selectivity for blocking the prejunctional α-adrenoceptors (Starke et al., 1975a), and other selective antagonists were identified (Borowski et al., 1977).

The designation of pre- and postjunctional α-adrenoceptors as $α_2$- and $α_1$-adrenoceptors, respectively, was proposed by Langer (1974). However, within a short period of time, evidence began to accumulate to suggest the presence of receptors similar to the autoreceptors at sites other than on adrenergic nerve terminals. Because a broad array of agonists and antagonists capable of discriminating between pre- and postjunctional α-adrenoceptors was available by that time, a pharmacological, rather than anatomical, subclassification of α-adrenoceptors into the $α_1$- and $α_2$-adrenoceptor subtypes was possible (Berthelsen and Pettinger, 1977). This subclassification was facilitated by the identification of prazosin as a highly selective competitive antagonist of postjunctional α-adrenoceptor (Doxey et al., 1977). As such, responses sensitive to blockade by prazosin were designated as $α_1$-adrenoceptor mediated, and those sensitive of blockade by yohimbine as $α_2$-adrenoceptor mediated (Starke, 1981).

Functional responses reflecting activation of pre- and postjunctional α-adrenoceptors in vivo were first demonstrated in the pithed rat (Drew, 1976), although earlier examples from the years before the discovery of $α_2$-autoreceptors can be found (Kobinger, 1967). Activation of prejunctional α-adrenoceptors would inhibit the tachycardia induced by spinal cord stimulation, whereas postjunctional α-adrenoceptor activation resulted in increases in blood pressure (Drew, 1976). The observations that the pressor responses to norepinephrine in the pithed rat and anesthetized cat were less sensitive to inhibition by prazosin than were the pressor responses to phenylephrine (Bentley et al., 1977; Drew and Whiting, 1979) led to the identification of postjunctional vascular $α_2$-adrenoceptors. By this time, several $α_2$-adrenoceptor agonists having greater selectivity than clonidine had been identified, including several azepines, such as B-HT 920 and B-HT 933, and aminotetralins, such as M-7. Rauwolscine, a diastereoisomer of yohimbine, had also been shown to have even greater selectivity for blocking $α_2$- versus $α_1$-adrenoceptors than yohimbine (Weitzell et al., 1979). Using these tools, several research groups convincingly demonstrated that activation of $α_2$-adrenoceptors could elevate blood pressure in the pithed rat, inasmuch as the pressor response to each $α_2$-adrenoceptor agonist was insensitive to inhibition by prazosin, but could be readily antagonized by yohimbine or rauwolscine (Hicks and Cannon, 1979; Timmermans *et al.*, 1979a; Timmermans *et al.*, 1979b; Docherty and McGrath, 1980; Kobinger and Pichler, 1980; Madjar *et al.*, 1980; Timmermans and van Zwieten, 1980).

At approximately the same time as postjunctional α_2-adrenoceptors were identified *in vivo* in the pithed rat, *in vitro* studies using the canine saphenous vein demonstrated the existence of postjunctional α_2-adrenoceptors which mediate vasoconstriction (DeMey and Vanhoutte, 1980; Shepperson and Langer, 1981). Rabbit (Schümann and Lues, 1983; Daly *et al.*, 1988a) and human (Docherty, 1987) saphenous veins were also shown to produce α_2-adrenoceptor mediated contraction, as did other veins (Daly *et al.*, 1988b). Although the pressor response to α_2-adrenoceptor agonists in the pithed rat clearly results from activation of arterial α_2-adrenoceptors, it was nonetheless difficult to demonstrate α_2-adrenoceptor mediated vasoconstriction in isolated arteries. More recently, α_2-adrenoceptor mediated vasoconstriction was demonstrated in small diameter arteries (Nielsen *et al.*, 1990). Furthermore, isolated arteries, such as canine saphenous artery (Sulpizio and Hieble, 1987) or rat tail artery (Craig *et al.*, 1995), were shown to contract in response to α_2-adrenoceptor activation, provided that threshold vasoconstriction was induced by agents opening calcium channels or activating eicosanoid receptors. Hence it is possible that endogenous vasoactive substances may facilitate the *in vivo* pressor responses to α_2-adrenoceptor agonists.

α_2-Adrenoceptors have been found in many locations and mediate a variety of functional responses. Prejunctional α_2-adrenoceptors are located on many peripheral and central neuronal terminals, including noradrenergic, cholinergic, and serotonergic, where their activation inhibits neurotransmitter release. α_2-Adrenoceptors also are present on neuronal cell bodies, where they mediate hyperpolarization and inhibition of firing rate. The adipocyte α_2-adrenoceptors, which mediate inhibition of lipolysis, have been characterized in several species, including hamster, dog and man (Schimmel, 1976; Taouis *et al.*, 1987; Hieble and Ruffolo, 1991). The platelet α_2-adrenoceptors that mediate aggregation have been studied primarily in human platelets, where they represent a useful model for functional and radioligand binding assays (Jakobs *et al.*, 1978; Grant and Scrutton, 1979). α_2-Adrenoceptors in the gastrointestinal mucosa inhibit fluid secretion. Other gastrointestinal α_2-adrenoceptors inhibit motility. Activation of α_2-adrenoceptors on pancreatic islet cells will inhibit insulin secretion. α_2-Adrenoceptors have also been identified as frog skin, where their activation inhibits α-MSH induced melanin granule dispersion (Pettinger, 1977). Activation of postjunctional α_2-adrenoceptors plays a role in the contractile response of the mouse vas deferens to norepinephrine or field stimulation sympathetic nerve terminals, and a contractile response to the selective α_2-adrenoceptor agonist, UK 14,304 can be demonstrated (Goncalves *et al.*, 1989; Bültmann *et al.*, 1991).

In some, but not all cases, an α_2-adrenoceptor antagonist will produce the opposite action from an agonist, suggesting that α_2-adrenoceptors may be tonically activated. For example, α_2-adrenoceptor antagonists will enhance norepinephrine overflow in sympathetically innervated tissues, both *in vitro* and *in vivo*, stimulate lipolysis, and potentiate glucose-induced insulin secretion.

α_2-Adrenoceptors mediate a variety of functions within the central nervous system, including reduction of sympathetic outflow, sedation and analgesia. Clonidine and α-methylnorepinephrine were shown to reduce blood pressure and sympathetic outflow by activation of α-adrenoceptors in the brainstem (Heise and Kroneberg, 1970; Schmitt *et al.*, 1971; Hoyer and van Zwieten, 1971). With the availability of subtype selective α-adrenoceptor antagonists, this response was shown to be mediated by α_2-adrenoceptor activation (Hamilton *et al.*, 1980; Timmermans *et al.*, 1981). The centrally mediated sympathoinhibitory action of clonidine appears to result from activation of α_2-adrenoceptors located on non-adrenergic nerves, since the ability of clonidine to decrease sympathetic outflow was not influenced by reserpine pretreatment (Haeusler, 1974). α_2-Adrenoceptor agonists depress firing rate in the locus coeruleus, an action similar to that produced by opiate receptor activation. α_2-Adrenoceptor activation can induce mydriasis, through actions within the brain and within the eye. The α_2-adrenoceptor is also linked to the control of intra-ocular pressure, although multiple effects are involved, and both stimulation and blockade of α_2-adrenoceptors can reduce intra-ocular pressure (see review by Potter, 1981).

α_2-Adrenoceptors can be detected using radioligand binding techniques in several tissues where functional responses to α_2-adrenoceptor agonists have not been detected, as is the case in the prostate gland and large arteries. These receptors may mediate responses that have not yet been characterized, or perhaps these receptors are not coupled to an effector system.

Radioligand binding assays

Concurrent with the identification of α_2-adrenoceptors in functional studies, the ability to study α-adrenoceptors using radioligand binding techniques in membrane homogenates was being developed. Several tritiated radioligands were initially used to label the α-adrenoceptors, including the endogenous catecholamines (U'Prichard and Snyder, 1977), clonidine (Greenberg *et al.*, 1976), WB-4101 (U'Prichard *et al.*, 1977), dihydroazepetine (Ruffolo *et al.*, 1976) and dihydroergokryptine (Peroutka *et al.*, 1978). The ligands studied most extensively were [^{3}H] clonidine, [^{3}H] WB-4101 and [^{3}H] dihydroergokryptine. [^{3}H] Dihydroergokryptine, a nonselective α-adrenoceptor antagonist, appeared to label the entire α-adrenoceptor population. Subtype selective antagonists were shown to produce a biphasic inhibition of [^{3}H] dihydroergokryptine binding suggesting the selective identification of α_1- and α_2-adrenoceptors in radioligand binding studies (Hoffman *et al.*, 1979). The relative affinities of the subtype selective antagonists for the two sites were used to quantitate α-adrenoceptor subtype selectivity, and the relative density of high and low affinity sites for a subtype selective antagonist, such as prazosin, was used to quantitate the distribution of α-adrenoceptor subtypes in a particular tissue.

Results obtained using [^{3}H] clonidine and [^{3}H] WB-4101 as radioligands were more complex to interpret. Most α-adrenoceptor agonists, regardless of their subtype selectivity in functional assays, were found to be more potent inhibitors of the binding of [^{3}H] clonidine than [^{3}H] WB-4101. Conversely, many antagonists were more potent inhibitors of the binding of [^{3}H] WB-4101. This led to the suggestion that [^{3}H] clonidine and [^{3}H] WB-4101 labeled agonist and antagonist states of the receptor, or distinct "agonist preferring" and "antagonist preferring" adrenoceptors (U'Prichard and Snyder, 1977; U'Prichard *et al.*, 1977).

However, the observation that selective α_2-adrenoceptor antagonists, such as yohimbine and rauwolscine were more potent inhibitors of [^{3}H] clonidine binding than [^{3}H] WB 4101 binding (U'Prichard *et al.*, 1977; Tanaka and Starke, 1980), led to the proposal that [^{3}H] clonidine selectively labeled α_2-adrenoceptors. This was subsequently confirmed by many laboratories over the past 15 years. [^{3}H] Clonidine remains a useful radioligand for evaluating α_2-adrenoceptors, although a variety of highly selective α_2-adrenoceptor antagonist radioligands are now available. [^{3}H] WB-4101 has been replaced as an α_1-adrenoceptor radioligand by the more selective antagonist ligands, [^{3}H] prazosin or [^{125}I] IBE-2254 (HEAT).

With the subtype selective radioligands currently available, agonists, such as norepinephrine, are still less potent inhibitors of the binding of antagonists than of [^{3}H] clonidine, even in membrane preparations containing pure populations of recombinant α_2-adrenoceptors. The ability of norepinephrine to inhibit the binding of antagonist radioligands is dependent on assay conditions, and in some cases, biphasic inhibition curves are obtained (Jansson et al., 1994; Kirifides and Codd, 1995). This phenomenon is thought to be a consequence of binding of the α_2-adrenoceptor to guanine nucleotide regulatory proteins that influence their affinity for agonists, since addition of GTP results in monophasic, low-affinity inhibition of binding (Jansson et al., 1994).

An α_2-adrenoceptor was purified to homogeneity in 1986 by affinity chromatography of human platelet membranes. The solubilized receptor had radioligand binding characteristics similar to those of the receptor in native platelet membranes (Regan et al., 1986). Using a similar technique, the α_2-adrenoceptors of porcine brain were isolated and purified (Repaske et al., 1987). Purification of α_2-adrenoceptor proteins from human platelet and neonatal rat lung provided evidence for physical and structural differences, consistent with the pharmacological differences between the α_2-adrenoceptor populations in these two tissues (see below) (Lanier et al., 1988).

TOOLS FOR PHARMACOLOGICAL CHARACTERIZATION OF THE α_2-ADRENOCEPTORS

Although both α_1- and α_2-adrenoceptors have now been further subdivided (see below), the use of selective agonists and antagonists to group adrenoceptor mediated responses into the two major classes remains a useful function.

Among agonists, methylation of epinephrine or norepinephrine at the α-carbon atom results in selectivity for the α_2- versus α_1-adrenoceptors (Ruffolo et al., 1982). Although α-methylnorepinephrine and α-methylepinephrine are full agonists at α_2-adrenoceptors, they retain significant agonist activity at both α_1- and β-adrenoceptors which in many cases limits their utility for receptor characterization.

In addition to clonidine, many imidazolines have been found to produce selective activation of the α_2-adrenoceptor. These include UK 14,304, which in many systems has greater intrinsic activity than clonidine, as well as analogs such as ST-91, SK&F 35886 and AGN 190851, which do not cross the blood brain barrier when administered systemically.

Other structurally diverse agents having selective α_2-adrenoceptor agonist activity include the azepines mentioned previously, B-HT 933 and B-HT 920, as well as the imidazoles, medetomidine and mivazerol.

The most commonly studied α_2-adrenoceptor antagonists are the yohimbine alkaloids. Rauwolscine remains the most useful compound in this series with respect to α_2- versus

α_1-adrenoceptor selectivity, but several other analogs retain substantial potency and selectivity for α_2-adrenoceptors (Weitzell *et al.*, 1979; McGrath, 1982). Several synthetic α_2-adrenoceptor antagonists are structural analogs of yohimbine, such as MK-912, WY-27127 and RS-15385-197. These compounds have even greater α_2- versus α_1-adrenoceptor selectivity than rauwolscine, and have lower affinity for other neurotransmitter receptors, such as dopamine and serotonin.

Many imidazolines are selective antagonists of α_2-adrenoceptors. One of the first members of this series to be identified was idazoxan (RX 781094). A minor structural modification of idazoxan results in another selective antagonist, RX 821002, which has even greater selectivity, and does not interact with the non-adrenergic sites which bind idazoxan and some other imidazolines (see Hieble and Ruffolo, 1995). Other selective α_2-adrenoceptor antagonists containing an imidazoline group include atipamezole and SL 84.0418.

A variety of cyclized phenethylamines are antagonists with selectivity for the α_2-adrenoceptors (Hieble *et al.*, 1985). A well characterized example from this group is the 3-benzazepine, SK&F 86466 (Hieble *et al.*, 1986).

SUBCLASSIFICATION OF THE α_2-ADRENOCEPTORS

Prejunctional α-adrenoceptors were the prototype α_2-adrenoceptors. Studies on prejunctional α-adrenoceptors also were the first to suggest that there is more than a single α_2-adrenoceptor (Doxey and Everitt, 1977; Dubocovich, 1979). Above all, a species difference between rat and rabbit α_2-adrenoceptors was a consistent finding: yohimbine and rauwolscine were less potent α_2-adrenoceptor antagonists than phentolamine and idazoxan in the rat, but were more potent antagonists than phentolamine and idazoxan in the rabbit (Taube *et al.*, 1977; Starke, 1981; Reichembacher *et al.*, 1982; Ennis, 1985; Lattimer and Rhodes, 1985; Alabaster *et al.*, 1986; Limberger *et al.*, 1989). The observations on rat and rabbit autoreceptors are now known to indeed represent the identification of distinct α_2-adrenoceptor subtypes. Ironically, it is a difference between orthologous receptors in the two species that was discovered in this manner: the α_2-adrenoceptors in the rat belong to the α_{2D} type (relatively low affinity for yohimbine and rauwolscine), whereas those in the rabbit belong to the orthologous α_{2A} type (relatively high affinity for yohimbine and rauwolscine) of a pari of orthologous receptors (see below).

Notwithstanding the pioneer role of the autoreceptors, the present α_2-adrenoceptor subclassification system is primarily based on radioligand binding data. Characterization of α_2-adrenoceptor binding in the neonatal rat lung showed an unexpectedly high affinity for prazosin, at that time thought to be a selective α_1-adrenoceptor antagonist, as an inhibitor of the binding of [^{3}H] yohimbine (Latifpour *et al.*, 1982; see also Cheung *et al.*, 1982). Relatively high affinity of prazosin against [^{3}H] yohimbine or [^{3}H] rauwolscine ($K_i < 100$ nM) was observed in other tissues, such rat kidney, as well as in some cell lines, such as NG-108 (Bylund, 1985; Nahorski *et al.*, 1985). In still other tissues, such as human platelet, or in the HT-29 cell line, the dissociation constant of prazosin against [^{3}H] rauwolscine was > 1000 nM (Bylund *et al.*, 1988; Bylund, 1992). The prazosin-insensitive α_2-adrenoceptor was designated as the α_{2A}-adrenoceptor, the prazosin-sensitive receptor as the α_{2B}-adrenoceptor (Bylund, 1985; Nahorski *et al.*, 1985). Analysis of data for the inhibition of [^{3}H] rauwolscine binding by prazosin demonstrated that other tissues, such as rat (Kawahara and Bylund, 1985) and human (Petrash and Bylund, 1986) cortex, had mixed populations of α_{2A}- and α_{2B}-adrenoceptors, indicating that this α_2-adrenoceptor subclassification did not simply represent a species difference in α_2-adrenoceptor characteristics. Other antagonists were subsequently identified that have selective actions at α_{2B}-adrenoceptors (ARC 239, spiroxatrine, SK&F 104856). Interestingly, most selective α_{2B}-adrenoceptor antagonists are also potent α_1-adrenoceptor antagonists, the only currently known exception being imiloxan (Michel et al., 1990).

In the subdivision of α_2-adrenoceptors via binding affinity, it was noted that the partial agonist, oxymetazoline, had the opposite selectivity profile to prazosin, interacting preferentially with the α_{2A}-adrenoceptor. Another imidazoline-containing molecule, BRL 44408, has been shown to be a selective α_{2A}-adrenoceptor antagonist (Young et al., 1989; Gleason and Hieble, 1992). BRL 48962, the R-enantiomer of BRL 44408, shows even greater selectivity for the α_{2a}-adrenoceptor[1], based on comparison of affinities for recombinant human α-adrenoceptors (Beeley et al., 1995).

Now that a variety of antagonists having a variety of selectivity profiles between the various α_2-adrenoceptor subtypes are available, a more precise mode of α_2-adrenoceptor characterization is possible, based on correlation of inhibitory potencies against α_2-adrenoceptor binding profiles for a series of antagonists between different receptor sources. This type of analysis led to the discovery of two additional α_2-adrenoceptor subtypes. The α_{2C}-adrenoceptor was initially shown to be present on a tissue culture cell line derived from the opossum kidney (Murphy and Bylund, 1988), and has also been found in native opossum kidney (Blaxall et al., 1991) as well as in a human retinoblastoma cell line (Gleason and Hieble, 1992). Studies using another α_2-adrenoceptor radioligand, [^{3}H] MK 912, have demonstrated a mixed population of α_2-adrenoceptors in rat cortex and spinal cord, with one component representing the α_{2C}-adrenoceptor subtype (Uhlén et al., 1992). Interestingly, this radioligand, like rauwolscine, appears to have a 10-fold greater affinity for α_{2C}-adrenoceptors, compared to α_{2B}-adrenoceptors. The differences between α_{2C}- and α_{2B}-adrenoceptors are subtle, with no one antagonist showing clear selectivity. The principal distinguishing characteristics of the α_{2C}-adrenoceptor is high affinity for rauwolscine, and a higher oxymetazoline/prazosin affinity ratio compared to that for the α_{2B}-adrenoceptor (Table 1). As can be seen from Table 1, the characteristics of these two subtypes are sufficiently close to make assignment ambiguous based only on the affinity of yohimbine, rauwolscine, prazosin and oxymetazoline. However, when the affinities of a more extensive series of antagonists are correlated, the differences between these subtypes become more apparent.

A fourth subtype, designated as the α_{2D}-adrenoceptor, has been found in bovine pineal and rat submaxillary gland (Simonneaux et al., 1991). This subtype has characteristics similar to the α_{2A}-adrenoceptor, but has a lower affinity for rauwolscine and yohimbine than the α_{2A}-adrenoceptor, resulting in a decreased yohimbine/prazosin affinity ratio (Table 1). The α_{2D}-adrenoceptor also has lower affinity for other alkaloids structurally analogous to yohimbine (O'Rourke et al., 1994). However, other α-adrenoceptor antago-

---

[1]As suggested for the α_1-adrenoceptors by the IUPHAR nomenclature committee (Hieble et al., 1995), native α_2-adrenoceptor subtypes are designated by upper case subscripts, and recombinant α_2-adrenoceptors with lower case subscripts.

TABLE 1

Radioligand binding affinities of antagonists to α_2-adrenoceptor subtypes

Source	K_i (nM)*				Affinity ratio**			
	YOH	RAU	PRZ	OXY	OXY/PRZ	YOH/OXY	YOH/PRZ	
Native								
Human Platelet	0.9	1.3	540	3.5	154	3.8	600	a
HT-29 Cell	1.8	1.2	2000	2.2	923	1.2	1128	b
Human Adipocyte	3	4	2200	16	137	5	733	Langin et al., 1990b
Canine Adipocyte	4		2700	16	158	4	675	Taouis et al., 1987
Recombinant								
α_{2a}-(Human)	3.2	4.5	1750	10.3	168	3.2	547	c
α_{2a}-(Porcine)	4.4		4100	21	197	5	941	Guyer et al., 1990
Native								
Neonatal Rat Lung	1	0.4	5	52	0.1	52	5	Michel et al., 1989
Rat Kidney	5	2.5	51	220	0.2	44	10	Bylund et al., 1988
NG-108 Cell	0.6	0.8	13	132	0.1	220	22	d
Recombinant								
α_{2b}-(Human)	6.0	4.1	110	1400	0.1	233	18	e
α_{2b}-(Rat)	8.7	9.6	46	610	0.1	70	5.3	f
α_{2b}-(Mouse)	12	7.3	59	1200	0.1	100	5	Chruschinski et al., 1992
Native								
Opossum Kidney	0.4	0.1	36	73	0.5	182	90	Blaxall et al., 1991
OK Cell	0.6	0.3	34	26	1.3	43	57	g
Y-79 Cell		0.4	123	14	8.8	35***	307***	Gleason and Hieble, 1992
Recombinant								
α_{2c}-(Human)	1.4	0.8	45	101	0.4	72	32	h
α_{2c}-(Mouse)	3.8	0.8	97	109	0.9	29	26	Link et al., 1992
α_{2c}-(Rat)	2.4	0.9	53	140	0.4	58	22	i
α_{2c}-(Opossum)	0.1	0.2		29			263	Blaxall et al., 1994a
$\alpha_{2?}$-(Fish)	1.1	0.3	469	49	9.5	45	426	Svensson et al., 1993

TABLE 1
Continued

Source	K_i (nM)[*]				Affinity ratio[**]			
	YOH	RAU	PRZ	OXY	OXY/PRZ	YOH/OXY	YOH/PRZ	
Native								
Hamster Adipocyte	33		2260	3	753	0.1	68	Saulnier-Blache et al., 1989 [a]
Rabbit Adipocyte	35		10000	14	642	0.4	285	Langin et al., 1990a [b]
Rat Adipocyte	70		1800	54	34	0.7	27	Carpene et al., 1990 [c]
Rat Enterocyte	54		1900	10	190	0.2	35	Paris et al., 1990 [d]
Rat Submaxillary		45	457	8	57	0.4[***]	25[***]	[j]
RINm5F Cell	104	18	1900	121	16	1.2	18	Remaury and Paris, 1992 [e]
Bovine Pineal	3.6	83	106	1.5	71	0.4	29	Simonneaux et al., 1991 [f]
Bovine Retina		12.1	3300	25	132	2.0[***]	272[***]	Bylund et al., 1995 [g]
Recombinant								
α_{2D}-(Rat)	59	57	1700	31	55	0.5	29	[k]
α_{2D}-(Mouse)	54	53	2150	33	65	0.6	40	Link et al., 1992

[*] K_i for inhibition of the binding of [^{3}H] yohimbine, [^{3}H] rauwolscine, [^{3}H] RX 821002 or [^{3}H] MK-912 to membrane homogenates.
[**] Affinity ratios represent the reciprocal of the ratio of K_i values
[***] rauwolscine/oxymetazoline or rauwolscine/prazosin

[a] Mean values from data reported by Bylund, 1992; Hieble and Naselsky, 1993; Brown et al., 1990; Daiguji et al., 1982.
[b] Mean values from data reported by Gleason and Hieble, 1991; Bylund et al., 1992; Langin et al., 1989.
[c] Mean values from data reported by Lanier et al., 1991, Lomasney et al., 1991; Bylund et al., 1992; Uhlén et al., 1994 and unpublished data from our laboratory.
[d] Mean values from Murphy and Bylund, 1988 and Gleason and Hieble, 1991.
[e] Mean values from Lomasney et al., 1991b; Bylund et al., 1992; Weinshank et al., 1990, Uhlén et al., 1994 and unpublished data from our laboratory.
[f] Mean values from Harrison et al., 1991 and Xia et al., 1993.
[g] Mean values from Bylund et al., 1992 and Gleason and Hieble, 1992.
[h] Mean values from Lomasney et al., 1990, Bylund et al., 1992 and Uhlén et al., 1994.
[i] Mean values from Harrison et al., 1991; Lanier et al., 1991; Flordellis et al., 1990; Voigt et al., 1991 and Uhlén et al., 1992.
[j] Mean values from Simonneaux et al., 1991 and Gleason and Hieble, 1992.
[k] Mean values from Harrison et al., 1991, Lanier et al., 1991 and Uhlén et al., 1993.

nists, such as RX 821002 and phentolamine, have affinities for the α_{2D}-adrenoceptor that are comparable to the other α_2-adrenoceptor subtypes. The dissociation constants for the binding of [^{3}H] RS-15385-197 to α_{2A}-, α_{2B}- and α_{2D}-adrenoceptors do not differ from one another (MacKinnon et al., 1992).

The subdivision of α_2-adrenoceptors into the four subtypes is based primarily on radioligand binding characteristics of the receptors in native tissue homogenates. However, as noted above, the discovery of subtypes in functional studies on prejunctional α_2-adrenoceptors preceded the radioligand-based classification. Recently, more quantitative analysis has compared the functional potency of antagonists at prejunctional α_2-adrenoceptors with their affinity for all four subtypes, as determined either in radioligand binding assays to native tissues possessing a single subtype, or to express recombinant α_2-adrenoceptors (see below). The results were of remarkable harmony: by far the majority of prejunctional α_2-autoreceptors and also α_2-heteroceptors belong either to the α_{2A}-adrenoceptor (rabbit atria, rabbit pulmonary artery, rabbit kidney, rabbit brain cortex and human saphenous vein as well as heteroceptors in rabbit brain cortex and caudate nucleus) or α_{2D}-adrenoceptor (rat submaxillary gland, rat vas deferens, rat kidney, rat brain cortex, mouse brain cortex, guinea pig brain cortex, guinea pig atria, guinea pig ileum as well as heteroceptors in guinea pig ileum) subtypes (Connaughton and Docherty, 1990; Schwartz and Malik, 1992; Smith and Docherty, 1992; Bohman et al., 1993; Trendelenburg et al., 1993; Funk et al., 1995; Limberger et al., 1995a,b; Molderings et al., 1995; Trendelenburg et al., 1995). The hypothesis was, therefore, put forward that prejunctional α_2-autoreceptors, and perhaps neuronal α_2-adrenoceptors in general, belong at least predominantly to the $\alpha_{2A/D}$ pair of orthologous receptors (see below; Trendelenberg et al., 1993; Smith et al., 1995; Starke et al., 1995). There are two apparent notable exceptions to this rule: the α_2-autoceptors in the human kidney and in human atria appear to be of the α_{2C} subtype (Trendelenburg et al., 1994; Rump et al., 1995). Although most of the prejunctional α_2-autoreceptors in the guinea pig appear to have α_{2D} characteristics, those of the urethra correspond better to the α_{2A}-adrenoceptor subtype (Alberts, 1992, 1995). This suggests that the guinea pig may possess both α_{2A}- and α_{2D}-adrenoceptors, which would not be in agreement with the premise (Bylund et al., 1995) that the α_{2a}- and α_{2d}-adrenoceptors represent species orthologs (see below).

While prejunctional α_2-adrenoceptors now are best classified functional α_2-adrenoceptors, there exist further functional counterparts for the α_{2A-D} subclassification. For example, ARC-239 is a more potent antagonist of α_2-adrenoceptor mediated inhibition of adenylate cyclase in NG-108 cells (α_{2B}-adrenoceptors) than in HT29 (α_{2A}) (Bylund and Ray-Prenger, 1989). In a similar study, prazosin was shown to be an effective antagonist of UK 14,304-induced inhibition of adenylate cyclase in OK cells, a response presumably mediated by α_{2C}-adrenoceptors (Murphy and Bylund, 1988). Although a dissociation constant for prazosin was not calculated in these experiments, prazosin appeared to be a less potent antagonist of the action of UK 14,304 in OK cells (relative to yohimbine) than in the NG-108 cells, consistent with binding affinity ratios at α_{2B}- and α_{2C}-adrenoceptors. Stimulation of prostaglandin synthesis in cultured smooth muscle cells from rabbit aorta also appears to be mediated by the α_{2C}-adrenoceptor (Nebigil and Malik, 1992). Nevertheless, most of the functional responses to α_2-adrenoceptor agonists in intact tissues appear to be mediated by α_{2A}- or α_{2D}-adrenoceptors. This is not unexpected, inasmuch as most of these responses were initially characterized by their sensitivity to rauwolscine and insensitivity to prazosin. A recent report (Craig *et al.*, 1995) suggests that UK 14,304-induced contraction of the rat caudal artery (in the presence of the thromboxane receptor agonist, U-46619) is mediated by the α_{2C}-adrenoceptor.

MOLECULAR BIOLOGY OF THE α_2-ADRENOCEPTORS

As noted above, the human platelet α_2-adrenoceptor protein was purified in 1986. An oligonucleotide probe prepared from a portion of this isolated protein was used to probe a human genomic library, and three distinct clones were found to hybridize to the probe (Kobilka *et al.*, 1988). One of these clones was expressed in *Xenopus oocytes*, and shown to have radioligand binding characteristics similar to those of the native platelet α_2-adrenoceptor. This receptor was localized to human chromosome 10, and hence was designated as α_2-C10. Subsequently, two other human α_2-adrenoceptors were cloned, expressed and characterized. Based on their chromosomal locations, these two additional α_2-adrenoceptors have been designated as α_2-C4 (Regan *et al.*, 1988) and α_2-C2 (Lomasney *et al.*, 1990; Weinshank *et al.*, 1990). Although there was initially some confusion with respect to the pharmacological relationships of these recombinant α_2-adrenoceptors to those identified in radioligand binding assays in native tissues, it is now clear that the α_2-C10, α_2-C2 and α_2-C4 adrenoceptors correspond to the α_{2A}-, α_{2B}- and α_{2C}-adrenoceptors, respectively (Table 1).

Three α_2-adrenoceptors have been cloned from the rat (Zeng *et al.*, 1990; Flordellis *et al.*, 1991; Harrison *et al.*, 1991; Lanier *et al.*, 1991; Voigt *et al.*, 1991), mouse (Chruschinski *et al.*, 1992; Link *et al.*, 1992) and guinea pig (Svensson *et al.*, 1995). Individual α_2-adrenoceptor subtypes have been cloned from the pig (α_{2a}; Guyer *et al.*, 1990) and opossum (α_{2c}; Blaxall *et al.*, 1994a). An α_2-adrenoceptor has been cloned from the skin of *Labrus ossifagus,* (Svensson *et al.*, 1993) a fish showing a melanocyte granule response to α_2-adrenoceptor stimulation similar to that observed by Pettinger (1977) in the frog. The α_{2b}- and α_{2c}-adrenoceptors appear to have similar pharmacological characteristics between species (Table 1). However, the α_{2a}-adrenoceptor cloned from human and porcine tissue has substantially higher affinity for yohimbine and rauwolscine than the homologous receptor cloned from rat, mouse or guinea pig (Table 1; Svensson *et al.*, 1995). The similar pharmacologic characteristics of the rat and mouse receptors to native tissues having the pharmacological profile of the α_{2D}-adrenoceptor has led to their designation as α_{2d} adrenoceptors, despite their high degree of amino acid identity (> 90%) to the human α_{2a}-adrenoceptor. The α_{2d}-adrenoceptor clones (rat, mouse, guinea pig) differ from the α_{2a}-adrenoceptor clones (human, porcine) in having a serine residue, rather than a cysteine, at the position corresponding to Cys^{201} in the α_{2a}-adrenoceptor. This change plays some role in the lower affinity observed for yohimbine, since mutation of Ser^{201} in the mouse α_{2D}-adrenoceptor to cysteine increases the affinity for yohimbine (Link *et al.*, 1995). However, other differences between α_{2d}- and α_{2a}-adrenoceptors must also influence antagonist affinity, since the above mutation does not substantially alter the affinity of the α_{2d}-adrenoceptor for rauwolscine and several other α_2-adrenoceptor antagonists (Blaxall *et al.*, 1994b).

Because only three α_2-adrenoceptors have been cloned to date from any single species (human, rat, guinea pig), it is now generally accepted that the α_{2A}- and α_{2D}-adrenoceptors are species orthologs, with a particular species having either α_{2A}- or α_{2D}-adrenoceptors, but not both (Bylund *et al.*, 1995). However, a few discrepancies from this premise remain to be resolved, such as the apparent existence of both α_{2A}- and α_{2D}-adrenoceptors at prejunctional sites in the guinea pig (see Funk *et al.*, 1995), and the proposed existence of α_{2D}-adrenoceptors in rabbit adipocytes (Langin *et al.*, 1990; Table 1), despite the functional classification of prejunctional α_2-adrenoceptors in the rabbit as being mediated by α_{2A}-adrenoceptors (Limberger *et al.*, 1995a; Molderings *et al.*, 1995).

CONCLUSIONS

It has been over 20 years since the discovery of α_2-adrenoceptors. During this time, much progress in their characterization has been made, with highly selective agonists and antagonists becoming available, multiple subtypes being identified, and the locations and functions of the α_2-adrenoceptors being assigned. The ability to delete selectively specific α_2-adrenoceptor subtypes from the mouse (Link *et al.*, 1995) may provide additional opportunities to explore their function. Several important goals remain, however, such as the identification of agonists and antagonists having greater selectivity for individual α_2-adrenoceptor subtypes, and the determination of functional roles for the α_{2B}- and α_{2C}-adrenoceptors. Furthermore, the existence of additional α_2-adrenoceptors, as yet uncharacterized by molecular biology, cannot be excluded.

Activation of the α_2-adrenoceptors has several potential therapeutic applications, including the treatment of hypertension, reduction of intra-ocular pressure, relief of the symptoms of opiate and ethanol withdrawal, and as adjuncts to general anesthesia (see reviews by Ruffolo *et al.*, 1993, 1995). Although the physiological roles of the α_2-adrenoceptors would suggest several therapeutic applications for α_2-adrenoceptor antagonists (*e.g.*, in the treatment of depression or non-insulin dependent diabetes), the clinical development of such agents has not yet been successful. The achievement of this goal, as well as the development of α_2-adrenoceptor agonists having an improved clinical profile with respect to clonidine and related drugs, may depend on further understanding of the roles of all of the α_2-adrenoceptor subtypes, and the development of agonists and antagonists having enhanced α_2-adrenoceptor subtype selectivity.

References

Alabaster, V.A., Keir, R.F. and Peters, C.J. (1986). Comparison of the potency of α_2-adrenoceptor antagonists in *in vitro*. Evidence for heterogeneity of α_2-adrenoceptors. *Br. J. Pharmacol.*, **88**, 607–615.

Alberts, P. (1992). Subtype classification of the presynaptic α-adrenoceptors which regulate [^{3}H]-noradrenaline secretion in guinea-pig isolated urethra. *Br. J. Pharmacol.*, **105**, 142–146.

Alberts, P. (1995). Presynaptic α_{2A}-adrenoceptors regulate the [^{3}H] noradrenaline secretion in the guinea pig urethra. *Pharmacol. and Toxicol.*, **77**, 95–101.

Beeley, L.J., Berge, J.M., Chapman, H., Hieble, P., Kelly, J., Naselsky, D.P., Rockell, C.M. and Young, P.W. (1995). Synthesis of a selective α_{2A} adrenocyptor antagonist, BRL 48962, and its characterization at cloned human α-adrenoceptors. *Bio-organic and Medical Chemistry*, in press.

Bentley, S.M., Drew, G.M. and Whiting, S.B. (1977). Evidence for two distinct types of postsynaptic α-adrenceptor. *Br. J. Pharmacol.*, **61**, 116P–117P.

Berthelsen, S. and Pettinger, W.A. (1977). A functional basis for the classification of α-adrenergic receptors. *Life Sci.*, **21**, 595–606.

Blaxall, H.S., Murphy, T.J., Baker, J.C., Ray, C. and Bylund, D.B. (1991). Characterization of the α_{2C} adrenergic receptor subtype in the opossum kidney and in the OK cell line. *J. Pharmacol. Exp. Ther.*, **259**, 323–329.

Blaxall, H.S., Cerutis, D.R., Hass, N.A., Iversen, L.J. and Bylund, D.B. (1994a). Cloning and expression of the α_{2C}-adrenergic receptor from the OK cell line. *Mol. Pharmacol.*, **45**, 176–181.

Blaxall, H.S., O'Rourke, M.F., Kobilka, B., Lomasney, J., Iversen, L.J., Heck, D.A. and Bylund, D.B. (1994b). Species characterization of $\alpha_{2A/D}$ adrenergic receptors. *Can. J. Physiol. Pharmacol.*, **72 (Suppl. 1)**, 542.

Bohmann, C., Schollmeyer, P. and Rump, L.C. (1993). α_2-Autoreceptor subclassification in rat isolated kidney by the use of short trains of electrical stimulation. *Br. J. Pharmacol.*, **108**, 262–268.

Borowski, E., Starke, K., Ehrl, H. and Endo, T. (1977). A comparison of pre- and postsynaptic effects of α-adrenolytic drugs in the pulmonary artery of the rabbit. *Neuroscience*, **2**, 285–296.

Brown, G.L. and Gillespie, J.S. (1957). The output of sympathetic transmitter from the spleen of the cat. *J. Physiol.* (London), **138**, 81–102.

Bültmann, R., von Kügelgen, I. and Starke, K. (1991). Contraction-mediating α_2-adrenoceptors in the mouse vas deferens. *Naunyn-Schmiedeberg's Arch. Pharmacol.*, **343**, 623–632.

Bylund, D.B. (1992). Subtypes of α_1 and α_2 adrenergic receptors. *FASEB J.*, **6**, 832–839.

Bylund, D.B. (1985) Heterogeneity of α_2-adrenergic receptors. *Pharmacol. Biochem. Behav.*, **22**, 835–843.

Bylund, D.B., Ray-Prenger, C. and Murphy, T.J. (1988). α_{2A} and α_{2B} Adrenergic receptor subtypes: antagonist binding in tissues and cell lines containing only one receptor subtype. *J. Pharmacol. Exp. Ther.*, **245**, 600–607.

Bylund, D.B. and Ray-Prenger, C. (1989). α_{2A} and α_{2B} adrenergic receptor subtypes: Attenuation of cyclic AMP production in cell lines containing only one receptor subtype. *J. Pharmacol. Exp. Ther.*, **251**, 640–644.

Bylund, D.B., Blaxall, H.S., Iversen, L.J., Caron, M.G., Lefkowitz, R.J. and Lomasney, J.W. (1992). Pharmacological characteristics of α_2-adrenergic receptors: Comparison of pharmacologically defined subtypes with subtypes identified by molecular cloning. *Mol. Pharmacol.*, **42**, 1–5.

Bylund, D.B., Regan, J.W., Faber, J.E., Hieble, J.P., Triggle, C.R. and Ruffolo, R.R., Jr. (1995). Vascular α-adrenoceptors: from the gene to the human. *Can. J. Physiol. Pharmacol.*, **73**, 533–543.

Cannon, W.B. and Bacq, Z.M. (1931). Studies on the conditions of activity in endocrine organs. XXII. A hormone produced by sympathetic action on smooth muscle. *Am. J. Physiol.*, **96**, 392–412.

Carpene, C., Galitzky, J., Larrouy, D., Langin, D. and LaFontan, M. (1990). Non-adrenergic sites for imdiazolines are not directly involved in the α_2-adrenergic antilipolytic effect of UK 14,304 in rat adipocytes. *Biochem. Pharmacol.*, **40**, 437–445.

Cheung, Y.-D., Barnett, D.B. and Nahorski, S.R. (1982). [^{3}H] Rauwolscine and [^{3}H] yohimbine binding to rat cerebral and human platelet membranes: Possible heterogeneity of α_2-adrenoceptors. *Europ. J. Pharmacol.*, **84**, 79–85.

Chruschinski, A.J., Link, R.E., Daunt, D.A., Barsh, G.S. and Kobilka, B.K. (1992). Cloning and expression of the mouse homolog of the human α_2-C2 adrenergic receptor. *Biochem. Biophys. Res. Com.*, **186**, 1280–1287.

Connaughton, S. and Docherty, J.R. (1990). Functional evidence for heterogeneity of peripheral prejunctional α_2-adrenoceptors. *Brit. J. Pharmacol.*, **101**, 285–290.

Craig, D., Iacolina, M. and Forray, C. (1995). α_{2C}-Adrenoceptors mediate norepinephrine-induced contraction of rat caudal artery. *FASEB J.*, **9**, A106.

Daiguji, M., Meltzer, H.Y. and U'Prichard, D.C. (1982). Human platelet α_2-adrenergic receptors: Labeling with ^{3}H-yohimbine, a selective antagonist ligand. *Life Sci.*, **28**, 2705–2717.

Daly, R.N., Sulpizio, A.C., Levitt, B., DeMarinis, R.M., Regan, J.W., Ruffolo, R.R., Jr. and Hieble, J.P. (1988a). Evidence for heterogeneity between pre- and postjunctional α_2 adrenoceptors using 9-substituted 3-benzazepines. *J. Pharmacol. Exp. Ther.*, **247**, 122–128.

Daly, C.J., McGrath, J.C. and Wilson, V.G. (1988b). Evidence that the population of postjunctional adrenoceptors mediating contraction of smooth muscle in the rabbit isolated ear vein is predominantly α_2. *Br. J. Pharmacol.*, **94**, 1085–1090.

DeMey, J.G. and Vanhoutte, P.M. (1980). Differences in pharmacological properties of postjunctional α-adrenergic receptors among arteries and veins. *Arch. Int. Pharmacodyn. Ther.*, **244**, 328–329.

Docherty, J.R. (1987). The use of the human saphenous vein in pharmacology. *Trends in Pharmacol. Sci.*, **8**, 358–361.

Docherty, J.R. and McGrath, J.C. (1980). A comparison of pre- and post-junctional potencies of several α-adrenoceptor agonists in the cardiovascular system and anococcygeus of the rat. Evidence for two types of postjunctional α-adrenoceptor. *Naunyn-Schmiedeberg's Arch. Pharmacol.*, **312**, 107.

Doxey, J.C., Smith, C.F.C. and Walker, J.M. (1977). Selectivity of blocking agents for pre- and postsynaptic α-adrenoceptors. *Br. J. Pharmacol.*, **60**, 91–96.

Drew, G.M. (1976). Effects of α-adrenoceptor agonists and antagonists on pre- and post-synaptically located α-adrenoceptors. *Eur. J. Pharmacol.*, **36**, 313–320.

Drew, G.M. and Whiting, S.B. (1979). Evidence for two distinct types of postsynaptic α-adrenoceptor in vascular smooth muscle *in vivo*. *Br. J. Pharmacol.*, **67**, 207–215.

Dubocovich, M.L. (1979). Pharmacological differences between the α-presynaptic adrenoceptors in the peripheral and the central nervous systems. In: Langer, S.Z., Starke, K. and Dubocovich, M.L., eds, *Presynaptic Receptors*. Pergamon Press, Oxford, pp. 29–36.

Dubocovich, M.L. and Langer, S.Z. (1974). Negative feed-back regulation of noradrenaline release by nerve stimulation in the perfused cat's spleen: differences in potency of phenoxybenzamine in blocking the pre- and postsynaptic adrenergic receptors. *J. Physiol.* (London), **237**, 505–519.

Ennis, C. (1985). Comparison of the α_2-adrenoceptors which modulate noradrenaline release in rabbits and rat occipital cortex. *Br. J. Pharmacol.*, **85**, 318P.

Farnebo, L.-O. and Hamberger, B. (1971). Drug-induced changes in the release of [^{3}H] monoamines from field-stimulated rat brain slices. *Acta Physiol. Scand.,* (**Suppl 371**), 35–44.

Flordellis, C.S., Handy, D.E., Bresnahan, M.R., Zannis, V.L. and Gavras, H. (1991). Cloning and expression of a rat brain α_{2b} adrenergic receptor. *Proc. Natl. Acad. Sci. USA*, **88**, 1019–1023.

Funk, L., Trendelenburg, A.-U., Limberger, N. and Starke, K. (1995). Subclassification of presynaptic α_2-adrenoceptors: α_{2D} adrenoceptors mediating release of acetylcholine in guinea pig ileum. *Naunyn-Schmiedeberg's Arch. Pharmacol.*, **352**, 58–66.

Gleason, M.M. and Hieble, J.P. (1991). Ability of SK&F 104078 and SK&F 104856 to identify α_2 adrenoceptor subtypes in NCB20 cells and guinea pig lung. *J. Pharmacol. Exp. Ther.*, **259**, 1124–1132.

Gleason, M.M. and Hieble, J.P. (1992). The α_2-adrenoceptors of the human retinoblastoma cell line (Y79) may represent an additional example of the α_{2C}-adrenoceptor. *Br. J. Pharmacol.*, **107**, 222–225.

Goncalves, J., Proenca, J., Albuquerque, A.A. and Paiva, M.Q. (1989). Hypothermia induces supersensitivity of mouse vas deferens to adrenergic agonists: evidence for postjunctional α_2-adrenoceptors. *J. Pharm. Pharmacol.*, **41**, 52–54.

Greenberg, D.A., U'Prichard, D.C. and Snyder, S.H. (1976). α-Noradrenergic receptor binding in mammalian brain: differential labelling of agonist and antagonist states. *Life Sci.*, **19**, 69–76.

Grant, J.A. and Scrutton, M.C. (1979). Novel α_2-adrenoceptors primarily responsible for inducing human platelet aggregation. *Nature,* **277**, 659–661.

Guyer, C.A., Horstman, D.A., Wilson, A.L., Clark, J.D., Cragoe, E.J. and Limbird, L.E. (1990). Cloning, sequencing and expression of the gene encoding the porcine α_2-adrenergic receptor. *J. Biol. Chem.*, **265**, 17307–17317.

Haeusler, G. (1974). Clonidine-induced inhibition of sympathetic nerve activity: No evidence for a central presynaptic action or an indirect sympathomimetic mode of action. *Naunyn-Schmiedeberg's Arch. Pharmacol.*, **286**, 97–111.

Hamilton, T.C., Hunt, A.E.E. and Poyser, R.H. (1980). Involvement of central α_2 adrenoceptors in the mediation of clonidine-induced hypotension in the cat. *J. Pharm. Pharmacol.*, **32**, 788–780.

Harrison, J.K., D'Angelo, D.D., Zeng, D. and Lynch, K.R. (1991). Pharmacological characterization of rat α_2-adrenergic receptors. *Mol. Pharmacol.*, **40**, 407–412.

Heise, A. and Kroneberg, G. (1970). Periphere und zentrale Kreislaufwirkung des α-Sympathicomimeticums 2-(2,6-Xylidino)-5,6-dihydro-4H-1,3-thiazinhydrochlorid (BAY 1470). *Naunyn-Schmiedeberg's Arch. Pharmacol.*, **266**, 350–351.

Hicks, P.E. and Cannon, J.G. (1979). N,N-Dialkyl derivatives of 2-amino-5,6-dihydroxy-1,2,3,4-tetrahydronaphthalene as selective agonists at presynaptic α-adrenoceptors in the rat. *J. Pharm. Pharmacol.*, **31**, 494–496.

Hieble, J.P., Bylund, D.B., Clarke, D.E., Eikenberg, D.C., Langer, S.Z., Lefkowitz, R.J, Minneman, K.P. and Ruffolo, R.R., Jr. (1995). International Union of Pharmacology Recommendation for nomenclature of α_1-adrenoceptors: *Consensus Update. Pharmacol. Rev.*, **47**, 267–270.

Hieble, J.P., Roesler, J.M., Fowler, P.J., Matthews, W.D. and DeMarinis, R.M. (1985). A new approach to antihypertensive therapy via blockade of the postjunctional α_2-adrenoceptor. In *Vascular Neuroeffector Mechanisms*, J.A. Bevan, T. Godfraind, R.A. Maxwell, J.C. Stcclet and M. Worcel, eds. Elsevier, Amsterdam, pp. 159–164.

Hieble, J.P. and Naselsky, D.P. (1993). Determination of α_2-adrenoceptor subtype selectivity in tissue homogenates and cell lines. *Pharmacologist,* **35**, 166.

Hieble, J.P., DeMarinis, R.M., Fowler, P.J. and Matthews, W.D. (1986). Selective α_2-adrenoceptor blockade by SK&F 86466: *In vitro* characterization of receptor selectivity. *J. Pharmacol. Exp. Ther.*, **236**, 90–96.

Hieble, J.P. and Ruffolo, R.R., Jr. (1991). Effects of α- and β-adrenoceptors on lipids and lipoproteins. In *Antilipidemic Drugs: Medicinal, Chemical and Biochemical Aspects*. D.T. Witiak, H.A.I. Newman and D.R. Feller, eds. Elsevier, Amsterdam, pp. 301–344.

Hieble, J.P. and Ruffolo, R.R., Jr. (1995). Possible structural and functional relationships between imidazoline receptors and α_2-adrenoceptors. *Ann. N.Y. Acad. Sci.*, **763**, 8–21.

Hoffman, B.B., de Lean, A., Wood, C.L., Schocken, D.D. and Lefkowitz, R.J. (1979). α-adrenergic receptor subtypes: Quantitative assessment by ligand binding. *Life Sci.*, **24**, 1739–1746.

Hoyer, I. and van Zwieten, P.A. (1971). The centrally induced fall in blood pressure after the infusion of amphetamine and related drugs into the vertebral artery of the cat. *J. Pharm. Pharmacol.*, **23**, 892–893.

Jakobs, K.H., Saur, W. and Schultz, G. (1978). Characterization of α- and β-adrenergic receptors linked to platelet adenylate cyclase. *Naunyn-Schmiedeberg's Arch Pharmacol.*, **302**, 285–291.

Jang, C-S. (1940). The potentiation and paralysis of adrenergic effects by ergotoxine and other substances. *J. Pharmacol.*, **71**, 87–94.

Jansson, C.C., Marjamaki, A., Luomala, K., Savola, J.M., Scheinin, M. and Akerman, K.E.O. (1994). Coupling of human α_2-adrenoceptor subtypes to regulation of cAMP production in transfected S115 cells. *Eur. J. Pharmacol. (Mol. Pharmacol.),* **266**, 165–174.

Kawahara, R.S. and Bylund, D.B. (1985). Solubilization and characterization of putative α_2-adrenergic isoceptors from the human platelet and the rat cerebral cortex. *J. Pharmacol. Exp. Ther.*, **233**, 603–610.

Kirifides, A.L. and Codd, E.E. (1995). High affinity agonist binding in transfected cloned human α_2-adrenoceptors. *Soc. of Neuroscience Abstracts*, in press.

Kirpekar, S.M. and Puig, M. (1971). Effect of flow-stop on noradrenaline release from normal spleens and spleens treated with cocaine, phentolamine or phenoxybenzamine. *Br. J. Pharmacol.*, **43**, 359–369.

Kobilka, B.K., Matsui, H., Kobilka, T.S., Yang-Feng, T.L., Francke, U., Caron, M.G., Lefkowitz, R.J. and Regan, J.W. (1988). Cloning, sequencing and expression of the gene coding for the human platelet α_2-adrenergic receptor. *Science*, **238**, 650–656.

Kobinger, W. (1967). Über den Wirkungsmechanismus einer neuen antihypertensiven Substanz mit Imidazolinstruktur. *Naunyn-Schmiedeberg's Arch., Pharmacol.,* **258**, 48–58.

Kobinger, W. and Pichler, L. (1980). Investigation into different types of post- and presynaptic α-adrenoceptors at cardiovascular sites in rats. *Eur. J. Pharmacol.*, **65**, 393–402.

Langer, S.Z., Adler, E., Enero, M.A. and Stefano, F.J.E. (1971). The role of the α-receptor in regulating noradrenaline overflow by nerve stimulation. *Proc. Intl. Union Physiol. Sci.*, **9**, 335.

Langer, S.Z. (1974). Presynaptic regulation of catecholamine release. *Brit. J. Pharmacol.*, **60**, 481–497.

Lanier, S.M., Homcy, C.J., Patenaude, C. and Graham, R.M. (1988). Identification of structurally distinct α_2-adrenergic receptors. *J. Biol. Chem.*, **263**, 14491–14496.

Lanier, S.M., Downing, S., Duzic, E. and Homcy, C.J. (1991). Isolation of rat genomic clones encoding subtypes of the α_2-adrenergic receptor. *J. Biol. Chem.*, **266**, 10470–10478.

Langin, D., LaFontan, M., Stillings, M.R. and Paris, S. (1989). [^{3}H]RX821002: A new tool for the identification of α_{2A}-adrenoceptors. *Eur. J. Pharmacol.*, **167**, 95–104.

Langin, D., Paris, H., Dauzats, M. and Lafontan, M. (1990a). Discrimination between α_2-adrenoceptors and [^{3}H]idazoxan-labelled non-adrenergic sites in rabbit white fat cells. *Eur. J. Pharmacol. (Mol. Pharmacology Sec.),* **188**, 261–272.

Langin, D., Paris, H. and Lafontan, M. (1990b). Binding of [^{3}H]idazoan and its methoxy derivative [^{3}H]RX821002 in human fat cells: [^{3}H]idazoxan, but not [^{3}H]RX821002 labels additional non-α_2-adrenergic binding sites. *Mol. Pharmacol.*, **37**, 876–885.

Latifpour, J., Jones, S.B. and Bylund, D.B. (1982). Characterization of [^{3}H] yohimbine binding to putative α_2-adrenergic receptors in neonatal rat lung. *J. Pharmacol. Exp. Ther.*, **223**, 606–611.

Lattimer, N. and Rhodes, K.F. (1985). A difference in the affinity of some selective α_2-adrenoceptor antagonists when compared on isolated vasa deferentia of rat and rabbit. *Naunyn-Schmiedeberg's Arch. Pharmacol.*, **329**, 278–281.

Limberger, N., Mayer, A., Zier, G., Valenta, B., Starke, K. and Singer, E.A. (1989). Estimation of pA$_2$ values at presynaptic α_2-autoreceptors in rabbit and rat brain cortex in the absence of autoinhibition. *Naunyn-Schmiedeberg's Arch. Pharmacol.*, **340**, 639–647.

Limberger, N., Funk, L., Trendelenburg, A.-U. and Starke, K. (1995a). Subclassification of presynaptic α_2-adrenoceptors: α_{2A}-autoreceptors in rabbit atria and kidney. *Naunyn-Schmiedeberg's Arch. Pharmacol.*, **352**, 31–42.

Limberger, N., Trendelenburg, A.-U. and Starke, K. (1995b). Subclassification of presynaptic α_2-adrenoceptors: α_{2D}-autoreceptors in mouse brain. *Naunyn-Schmiedeberg's Arch. Pharmacol.*, **352**, 43–48.

Link, R., Daunt, D., Barsh, G., Chruscinski, A. and Kobilka, B. (1992). Cloning of two mouse genes encoding α_2-adrenergic receptor subtypes and identification of a single amino acid in the mouse α_2-C10 homolog responsible for an interspecies variation in antagonist binding. *Mol. Pharmacol.*, **42**, 16–27.

Link, R.E., Stevens, M.S., Kulatunga, M., Scheinin, M., Barsh, G.S. and Kobilka, B.K. (1995). Targeted inactivation of the gene encoding the mouse α_{2C} adrenoceptor homolog. *Mol. Pharmacol.*, **48**, 48–55.

Lomasney, J.W., Lorenz, W., Allen, L.F., King, K., Regan, J.W., Yang-Feng, T.L., Caron, M.G. and Lefkowitz, R.J. (1990). Expansion of the α_2-adrenergic receptor family: Cloning and expression of a human α_2-adrenergic receptor subtype, the gene for which is located on chromosome 2. *Proc. Natl. Acad. Sci. USA*, **87**, 5094–5098.

Lomasney, J.W., Cotecchia, S., Lefkowitz, R.J. and Caron, M.G. (1991). Molecular biology of α-adrenergic receptors: Implications for receptor classification and for structure-function relationships. *Biochim. et Biophys. Acta*, **1095**, 127–139.

MacKinnon, A.C., Kilpatrick, A.T., Kenny, B.A., Spedding, M. and Brown, C.M. (1992). [^{3}H]RS-15385-197, a selective and high affinity radioligand for α_2-adrenoceptors: Implications for receptor classification. *Br. J. Pharmacol.*, **106**, 1011–1018.

Madjar, H., Docherty, J.R. and Starke, K. (1980). An examination of pre- and postsynaptic α-adrenoceptors in the autoperfused rabbit hindlimb. *J. Cardiovasc. Pharmacol.*, **2**, 619–627.

McGrath, J.C. (1982). Evidence for more than one type of post-junctional α-adrenoceptor. *Biochem. Pharmacol.*, **31**, 467–484.

Michel, A.D., Loury, D.N. and Whiting, R.L. (1989). Differences between the α_2-adrenoceptor in rat submaxillary gland and the α_{2A}- and α_{2B}-adrenoceptor subtypes. *Brit. J. Pharmacol.*, **98**, 890–897.

Michel, A.D., Loury, D.N. and Whiting, R.L. (1990). Assessment of imiloxan as a selective α_{2B}-adrenoceptor antagonist. *Br. J. Pharmacol.*, **99**, 560–564.

Molderings, G. and Gothert, M. (1995). Subtype determination of presynaptic α_2-autoreceptors in the rabbit pulmonary artery and human saphenous vein. *Naunyn-Schmiedeberg's Arch. Pharmacol.*, in press.

Murphy, T.J. and Bylund, D.B. (1988). Characterization of α_2 adrenergic receptors in the OK cell, an opossum kidney cell line. *J. Pharmacol. Exp. Ther.*, **244**, 571–578.

Nahorski, S.R., Barnett, D.B. and Cheung, Y.-D. (1985). α-Adrenoceptor-effector coupling: affinity states or heterogeneity of the α_2-adrenoceptor? *Clin. Sci.*, **68 (Suppl. 10)**, 39s–42s.

Nebigil, C. and Malik, K.U. (1992). Comparison of signal transduction mechanisms of α_{2C} and α_{1A} adrenergic receptor stimulated prostaglandin synthesis. *J. Pharmacol. Ther.*, **263**, 987–996.

Nielsen, H., Mortensen, F.V. and Mulvany, M.J. (1990). Differential distribution of postjunctional α_2-adrenoceptors in human omental small arteries. *J. Cardiovasc. Pharmacol.*, **16**, 34–40.

O'Rourke, M.F., Iversen, L.J., Lomasney, J.W. and Bylund, D.B. (1994). Species orthologs of the α_{2A} adrenergic receptor: The pharmacological properties of the bovine and rat receptors differ from the human and porcine receptors. *J. Pharmacol. Exp. Ther.*, **271**, 735–740.

Paris, H., Voisin, T., Remaury, A., Rouyer-Fessard, C., Daviaud, D. and Langin, D. (1990). α_2 adrenoceptor in rat jejunum epithelial cells: Characterization with [^{3}H] RX 821002 and distribution along the villus-crypt axis. *J. Pharmacol. Exp. Ther.*, **254**, 888–893.

Peroutka, S.J., Greenberg, D.A., U'Prichard, D.C. and Snyder, S.H. (1978). Regional variation in α adrenergic receptor interactions of [^{3}H] dihydroergokryptine in calf brain: implications for a two-site model of α receptor function. *Mol. Pharmacol.*, **14**, 403–412.

Petrash, A.C. and Bylund, D.B. (1986). α_2 adrenergic receptor subtypes indicated by [^{3}H] yohimbine binding in human brain. *Life Sci.*, **38**, 2129–2137.

Pettinger, W.A. (1977). Unusual α-adrenergic receptor potency of methyldopa metabolites on melanocyte function. *J. Pharmacol. Exp. Ther.*, **201**, 622–626.

Potter, D.E. (1981). Adrenergic pharmacology of aqueous humor dynamics. *Pharmacol. Rev.*, **33**, 133–153.

Regan, J.W., Nakata, H., DeMarinis, R.M., Caron, M.G. and Lefkowitz, R.J. (1986). Purification and characterization of the human platelet α_2-adrenergic receptor. *J. Biol. Chem.*, **261**, 3894–3900.

Regan, J.W., Kobilka, T.S., Yang-Feng, T.L., Caron, M.G., Lefkowitz, R.J. and Kobilka, B.K. (1988). Cloning and expression of a human kidney cDNA for an α_2-adrenergic receptor subtype. *Proc. Natl. Acad. Sci. USA*, **85**, 6301–6305.

Reichenbacher, D., Reimann, W. and Starke, K. (1982). α-Adrenoceptor-mediated inhibition of noradrenaline release in rabbit brain slices. *Naunyn-Schmiedeberg's Arch. Pharmacol.*, **319**, 71–77.

Remaury, A. and Paris, H. (1992). The insulin secreting celll line, RINm5F, expresses an α_{2D} adrenoceptor and nonadrenergic idazoxan binding sites. *J. Pharmacol. Exp. Ther.*, **260**, 417–426.

Repaske, M.G., Nunnari, J.M. and Limbird, L.E. (1987). Purification of the α_2-adrenergic receptor from porcine brain using a yohimbine-agarose affinity matrix. *J. Biol. Chem.*, **262**, 12381–12386.

Ruffolo, R.R., Jr., Fowble, J.W., Miller, D.D. and Patil, P.N. (1976). Binding of ^{3}H-dihydroazapetine to α-adrenoreceptor-related proteins from rat vas deferens. *Proc. Natl. Acad. Sci. USA*, **73**, 2730–2734.

Ruffolo, R.R., Jr., Yaden, E.L. and Waddell, J.E. (1982). Stereochemical requirements of α_2 adrenergic receptors. *J. Pharmacol. Exp. Ther.*, **222**, 645–651.

Ruffolo, R.R., Jr., Nichols, A.J., Stadel, J.M. and Hieble, J.P. (1993). Pharmacologic and therapeutic applications of α_2-adrenoceptor subtypes. *Ann. Rev. Pharmacol. Toxicol.*, **32**, 243–279.

Ruffolo, R.R., Jr., Bondinell, W. and Hieble, J.P. (1995). α- and β-adrenoceptors: From the gene to the clinic. Part II. Structure-activity relationships and therapeutic applications. *J. Med. Chem.*, **38**, 3681–3716.

Rump, C.L., Bohmann, C., Schauble, U., Schöllhorn, J. and Limberger, N. (1995). α_{2C}-Adrenoceptor-modulated release of noradrenaline in human right atrium. *Brit. J. Pharmacol.*, in press.

Saulnier-Blache, J., Carpene, C., Langin, D. and LaFontan, M. (1989). Imidazolinic radioligands for the identification of hamster adipocyte α_2-adrenoceptors. *Eur. J. Pharmacol.*, **171**, 145–157.

Schimmel, R.J. (1976). Roles of α- and β-adrenoceptors in the control of glucose oxidation in hamster epididymal adipocytes. *Biochim. Biophys. Acta*, **428**, 379–387.

Schmitt, H., Schmitt, H. and Fenard, S. (1971). Evidence for an α-sympathomimetic component in the effects of catapresan on vasomotor centres: antagonism by piperoxane. *Eur. J. Pharmacol.*, **14**, 98–100.

Schümann, H.J. and Lues, I. (1983). Postjunctional α-adrenoceptors in the isolated saphenous vein of the rabbit. Characterization and influence of angiotensin. *Naunyn-Schmiedeberg's Arch. Pharmacol.*, **323**, 328–334.

Schwartz, D.D. and Malik, K.U. (1992). Characterization of prejunctional α_2 adrenergic receptors involved in modulation of adrenergic transmitter release in the isolated perfused rat kidney. *J. Pharmacol. Exp. Ther.*, **261**, 1050–1055.

Shepperson, N.B. and Langer, S.Z. (1981). The effects of the 2-amino-tetrahydro-naphthalene derivative M7, a selective α_2-adrenoceptor agonist *in vitro*. *Naunyn-Schmiedeberg's Arch. Pharmacol.*, **318**, 10–13.

Simonneaux, V., Ebadi, M. and Bylund, D.B. (1991). Identification and characterization of α_{2D}-adrenergic receptors in bovine pineal gland. *Mol. Pharmacol.*, **40**, 235–241.

Smith, K. and Docherty, J.R. (1992). Are the prejunctional α_2-adrenoceptors of the rat vas deferens and submandibular gland of the α_{2A} or α_{2D} subtype? *Europ. J. Pharmacol.*, **219**, 203–210.

Smith, K., Gavin, K. and Docherty, J.R. (1995). Investigation of the subtype of α_2-adrenoceptor mediating prejunctional inhibition of cardioacceleration in the pithed rat heart. *Brit. J. Pharmacol.*, **115**, 316–320.

Starke, K. (1971). Influence of α-receptor stimulants on noradrenaline release. *Naturwissenschaften*, **58**, 420

Starke, K. (1972a). A sympathomimetic inhibition of adrenergic and cholinergic transmission in the rabbit heart. *Naunyn-Schmiedeberg's Arch. Pharmacol.*, **274**, 18–45.

Starke, K. (1972b). Influence of extracellular noradrenaline on the stimulation-evoked secretion of noradrenaline from sympathetic nerves: evidence for an α-receptor mediated feed-back inhibition of noradrenaline release. *Naunyn-Schmiedeberg's Arch. Pharmacol.*, **275**, 11–23.

Starke, K. (1977). Regulation of noradrenaline release by presynaptic receptor systems. *Rev. Physiol. Biochem. Pharmacol.*, **77**, 1–124.

Starke, K. (1981). α-Adrenoceptor subclassification. *Rev. Physiol. Biochem. Pharmacol.*, **88**, 199–236.

Starke, K., Montel, H., and Schümann, H.J. (1971a). Influence of cocaine and phenoxybenzamine on noradrenaline uptake and release. *Naunyn-Schmiedeberg's Arch. Pharmacol.*, **270**, 210–214.

Starke, K., Montel, H. and Wagner, J. (1971b). Effect of phentolamine on noradrenaline uptake and release. *Naunyn-Schmiedeberg's Arch. Pharmacol.*, **271**, 181–192.

Starke, K., Montel, H., Gayk, W. and Merker, R. (1974). Comparision of the effects of clonidine on pre- and postsynaptic adrenoceptors in the rabbit pulmonary artery. *Naunyn-Schmiedeberg's Arch. Pharmacol.*, **285**, 133–150.

Starke, K., Borowski, E. and Endo, T. (1975a). Preferential blockade of presynaptic α-adrenoceptors by yohimbine. *Eur. J. Pharmacol.*, **34**, 385–388.

Starke, K., Endo, T. and Taube, H.D. (1975b). Relative pre- and postsynaptic potencies of α-adrenoceptor agonists in the rabbit pulmonary artery. *Naunyn-Schmiedeberg's Arch. Pharmacol.*, **291**, 55–78.

Steinsland, O.S. and Nelson, S.H. (1975). "α-adrenergic" inhibition of the response of the isolated rabbit ear artery to brief intermittent sympathetic nerve stimulation. *Blood Vessels*, **12**, 378–379.

Sulpizio, A.C. and Hieble, J.P. (1987). Demonstration of α_2-adrenoceptor-mediated contraction in the isolated canine saphenous artery treated with Bay K 8644. *Eur. J. Pharmacol.*, **135**, 107–110.

Svensson, S.P.S., Bailey, T.J., Pepperl, D.J., Grundstrom, N., Ala-Uotila, S., Scheinin, M., Karlsson, J.O.G. and Regan, J.W. (1993). Cloning and expression of a fish α_2-adrenoceptor. *Br. J. Pharmacol.*, **110**, 54–60.

Svensson, S.P.S., Bailey, T.J., Porter, A.C., Richman, J.G. and Regan, J.W. (1995). Heterologous expression of the cloned guinea pig α_{2a}, α_{2b} and α_{2c} adrenoceptor subtypes: Radioligand binding and functional coupling to a cAMP responsive reporter gene. *Biochem. Pharmacol.*, in press.

Tanaka, T. and Starke, K. (1980). Antagonist/agonist preferring α-adrenoceptors or α_1/α_2 adrenoceptors. *Eur. J. Pharmacol.*, **63**, 191–194.

Taouis, M., Berlan, M., Mantastruc, P. and LaFontan, M. (1987). Characterization of dog fat cell α_2-adrenoceptors: Variations in α_2 and β-adrenergic receptor distribution according to the extent of the fat deposits and anatomical location. *J. Pharmacol. Exp. Ther.*, **242**, 1041–1049.

Taube, H.D, Starke, K. and Borowski, E. (1977). Presynaptic receptor systems on the noradrenergic neurones of rat brain. *Naunyn-Schmiedeberg's Arch. Pharmacol.*, **299**, 123–141.

Timmermans, P.B.M.W.M., Kwa, H.Y. and van Zwieten, P.A. (1979a). Possible subdivision of postsynaptic α-adrenoceptors mediating pressor responses in the pithed rat. *Naunyn-Schmiedeberg's Arch. Pharmacol.*, **310**, 189–193.

Timmermans, P.B.M.W.M., Lam, E. and van Zwieten, P.A. (1979b). The interaction between prazosin and clonidine at α-adrenoceptors in rats and cats. *Eur. J. Pharmacol.*, **55**, 57–66.

Timmermans, P.B.M.W.M., Schoop, A.M.C., Kwa, H.Y. and van Zwieten, P.A. (1981). Characterization of α-adrenoceptors participating in the central hypotensive and sedative effects of clonidine using yohimbine, rauwolscine and corynanthine. *Eur. J. Pharmacol.*, **70**, 7–15.

Timmermans, P.B.M.W.M. and van Zwieten, P.A. (1980). Postsynaptic α_1 and α_2 adrenoceptors in the circulatory system of the pithed rat: Selective stimulation of the α_2 type by B-HT 933. *Eur. J. Pharmacol.*, **63**, 199–202.

Trendelenburg, A.U., Limberger, N. and Rump, L.C. (1994). α_2-Adrenergic receptors of the α_{2C} subtype mediate inhibition of norepinephrine release in human kidney cortex. *Mol. Pharmacol.*, **45**, 1168–1176.

Trendelenburg, A.U., Limberger, N. and Starke, K. (1993). Presynaptic α_2-autoreceptors in brain cortex: α_{2D} in the rat and α_{2A} in the rabbit. *Naunyn-Schmiedeberg's Arch. Pharmacol.*, **348**, 35–45.

Trendelenburg, A.U., Limberger, N. and Starke, K. (1995). Subclassification of presynaptic α_2-adrenoceptors: α_{2D}-autoreceptors in guinea-pig atria and brain. *Naunyn-Schmiedeberg's Arch. Pharmacol.*, **352**, 49–57.

Uhlén, S., Xia, Y., Chhajlani, V., Lien, E.J. and Wikberg, J.E.S. (1993). Evidence for the existence of two forms of α_{2A} adrenoceptors in the rat. *Naunyn-Schmiedeberg's Arch. Pharmacol.*, **347**, 280–288.

Uhlén, S., Porter, A.C. and Neubig, R.R. (1994). The novel α_2 adrenergic radioligand [^{3}H] -MK912 is α_{2C} selective among human α_{2A}, α_{2B} and α_{2C} adrenoceptors. *J. Pharmacol. Exp. Ther.*, **271**, 1558–1565.

Uhlén, S., Xia, Y., Chhajlani, V., Felder, C.C. and Wikberg, J.E.S. (1992). [^{3}H]-MK 912 binding delineates two α_2-adrenoceptor subtypes in rat CNS one of which is identical with the cloned pA$_{2d}$ α_2-adrenoceptor. *Br. J. Pharmacol.*, **106**, 986–995.

U'Prichard, D.C. and Snyder, S.H. (1977). Binding of ^{3}H catecholamines to α-noradrenergic receptor sites in calf brain. *J. Biol. Chem.*, **252**, 6450–6463.

U'Prichard, D.C., Greenberg, D.A. and Snyder, S.H. (1977). Binding characteristics of a radiolabeled agonist and antagonist at central nervous system α noradrenergic receptors. *Mol. Pharmacol.*, **13**, 454–473.

Voigt, M.M., McCune, S.K., Kanterman, R.Y. and Felder, C.C. (1991). The rat α_2-C4 adrenergic receptor gene encodes a novel pharmacological subtype. *FEBS Letters*, **278**, 45–50.

Weinshank, R.L., Zgombick, J.M., Macchi, M., Adham, N., Lichtblau, H., Branchek, T.A. and Hartig, P.A. (1990). Cloning, expression and pharmacological characterization of a human α_{2B}-adrenergic receptor. *Mol. Pharmacol.*, **38**, 681–688.

Weitzell, R., Tanaka, T. and Starke, K. (1979). Pre-postsynaptic effects of yohimbine stereoisomers on noradrenergic transmission in the pulmonary artery of the rabbit. *Naunyn-Schmiedeberg's Arch. Pharmacol.*, **308**, 127–136.

Werner, U., Starke, K. and Schümann, H.J. (1970). Wirkungen von Clonidin (ST 155) and BAY a 6781 auf das isolierte Kaninchenherz. *Naunyn-Schmiedeberg's Arch Pharmacol.*, **266**, 474.

Xia, Y., Uhlén, S., Chhajlani, V., Lien, E.J. and Wikberg, J.E.S. (1993). Further evidence for the existence of two forms of α_{2B}-adrenoceptors in rat. *Pharmacol. Toxicol.*, **72**, 40–49.

Young, P., Berge, J., Chapman, H. and Cawthorne, M.A. (1989). Novel α_2-adrenoceptor antagonists show selectivity for α_{2A}- and α_{2B}-adrenoceptor subtypes. *Eur. J. Pharmacol.*, **168**, 381–386.

Zeng, D., Harrison, J.K., D'Angelo, D.D., Barber, C.M., Tucker, A.L., Lu, Z. and Lynch, K.R. (1990). Molecular characterization of a rat α_{2B}-adrenergic receptor. *Proc. Natl. Acad. Sci. USA*, **87**, 3102–3106.

α_{2a} ADRENERGIC RECEPTOR PEPTIDES AND G PROTEINS: STRUCTURE AND MECHANISM

RICHARD R. NEUBIG,[1,2] JOAN M. TAYLOR,[1] SUSAN M. WADE[1] and SHING-ZHAO YANG[1]

Departments of Pharmacology[1] and Internal Medicine[2], University of Michigan, Ann Arbor, MI 48109

The regions of α_2 adrenergic receptors which are involved in coupling to G proteins were examined by use of synthetic peptide fragments which mimic receptor function. Structure function studies were done with a peptide (Q) from the carboxyterminal end of the third intracellular loop (i3c) of the porcine α_{2A} adrenergic receptor. The data are consistent with a model in which a basic region approximately 15 to 18 residues from the 6th transmembrane span is an activator of G protein while the region nearer the membrane regulates the contact of this "effector" region with the G protein. The i2 and i3n regions thus may be more important for specificity of G protein coupling. Sites on the G protein were also probed by a photoaffinity derivative of this peptide (DAP-Q). The photoprobe incorporates into the extreme amino-terminal region of G_o α subunit and in the carboxy-terminal region of Gβ. A functional map and hypothetical model of G$\beta\gamma$ structure is presented to account for its organization in the membrane and receptor contact sites.

KEY WORDS: AR – Adrenergic receptor, DAP – Diazopyruvoyl, DAP-Q – DAP modified Q peptide, G_o – Abundant G protein purified from bovine brain, G_i – Inhibitory G protein purified from bovine brain, GPCR – G protein coupled receptor, i3 – Third intracellular loop, Q peptide – Peptide with the sequence RWRGRQNREKRFTC (aa 361-373). from the porcine α_{2A}-adrenergic receptor with additional carboxy-terminal cysteine, TID – 3-trifluoromethyl-3-(*m*-iodophenyl)diazirine

INTRODUCTION

Heterotrimeric guanine nucleotide binding proteins (G proteins) transduce many different signals in mammalian cells (Brown and Birnbaumer, 1990; Hepler and Gilman, 1992). While the general regions of receptors involved in G protein coupling are known (Kobilka *et al.*, 1988; Strader *et al.*, 1989), the structural and mechanistic basis for G protein activation is much less clear. Synthetic peptides derived from the sequence of biologically active proteins are also useful as tools to study protein-protein interactions and as potential therapeutic agents. Peptides from signal transducing molecules such as receptors and heterotrimeric G proteins (Malek *et al.*, 1993; Neubig and Dalman, 1991; Dalman *et al.*, 1991; Ikezu *et al.*, 1992; Dalman and Neubig, 1991; Munch *et al.*, 1991; Taylor and Neubig, 1994; Hargrave *et al.*, 1993; König *et al.*, 1989; Hamm *et al.*, 1988) have been used for these purposes.

Correspondence to: Richard Neubig, Dept. of Pharmacology, University of Michigan, 1301 MSRBIII, Ann Arbor, MI 48109-0632. Phone: (313) 763-3650   Fax: (313) 763-4450   E-mail: RNeubig@umich.edu

Mastoparan a tetradecapeptide from wasp venom activates signals in mast cells and was one of the first small peptides found to have direct G protein activating activity (Higashijima et al., 1990). Also, peptides from the retinal G protein transducin (Hamm et al., 1988) and from three intracellular regions of its cognate receptor, rhodopsin, (König et al., 1989) were shown to disrupt rhodopsin-transducin interactions. Subsequently, synthetic peptides from numerous receptors including the β adrenergic (Munch et al., 1991), α_2 adrenergic (Wagner et al., 1995; Dalman et al., 1991; Ikezu et al., 1992; Dalman and Neubig, 1991), dopamine D_2 (Malek et al., 1993) and 5HT1a (Varrault et al., 1994) receptors have been shown to affect receptor and G protein activity. Structural features such as amphiphilicity (Higashijima et al., 1990). and a "consensus sequence" BBXB (Okamoto and Nishimoto, 1992) have been proposed to be important but contrary views have been expressed (Wade et al., 1994; Voss et al., 1993). Basic and hydrophobic residues also appear to play a role in RG coupling (Higashijima et al., 1990; Mukai et al., 1992; Ikezu et al., 1992).

A peptide from the carboxyl-terminal portion of the third intracellular loop (i3c) of the α_2 adrenergic receptor has proved to be a very useful probe of the α_2 receptor contact sites on pertussis toxin sensitive G proteins. It disrupts high affinity agonist binding and epinephrine-stimulated GTPase stimulation in human platelet membranes (Dalman and Neubig, 1991); it behaves as a weak partial agonist for G proteins by directly activating purifed G_o and G_i (Wade et al., 1994); a photoaffinity derivative binds to a specific site on the N-terminus of the G_o α subunit (Taylor et al., 1994b); and as will be discussed here it also interacts with a defined site on the β subunit (Taylor et al., 1996; Taylor et al., 1994b). To begin to define critical molecular determinants of receptor-mediated G protein activation we describe structure-activity studies of the α_{2a} adrenergic receptor i3c peptide and mapping of its sites of cross-linking to G protein subunits. Because the role of Gβ subunits in functional interactions is poorly defined, we also review the literature on functional domains of the Gβ subunit.

MATERIALS AND METHODS

Peptide synthesis. Peptides P and Q (Figure 1) with sequences from the third intracellular loop of the porcine α_2 adrenergic receptor (Guyer et al., 1990) were synthesized by the University of Michigan Protein and Carbohydrate Structure Core Facility using Fmoc (fluorenylmethoxycarbonyl) chemistry as described (Dalman and Neubig, 1991). P peptide (CRIYQIAKRRTRV) is from the amino-terminal end of the third intracellular loop (i3n) and Q peptide (RWRGRQNREKRFTC) is from the carboxyl-terminal end of the i3 loop (i3c) of the porcine α_{2a} adrenergic receptor. Both have a cysteine residue in addition to the native receptor sequence. The Q-Thr^{373} substituted peptides and the tetradecapeptide mastoparan from wasp venom (INLKALAALAKKIL) were also synthesized by the University of Michigan Protein and Carbohydrate Structure Core Facility. All of these peptides were evaluated for purity by HPLC and electrospray mass spectrometry.

A series of Q peptide analogs truncated from each end (Wade et al., 1996) was synthesized by Cambridge Research Biochemicals (Wilmington, DE). They were further purified by reverse phase HPLC with either an analytical (0.46 × 25 cm) column or semi-preparative (1.0 × 25 cm) Vydac 218TP C_{18} column eluted with 0.1% TFA and an acetonitrile gradient. Isolated peaks were collected and dried using a Savant SS1 SpeedVac system.

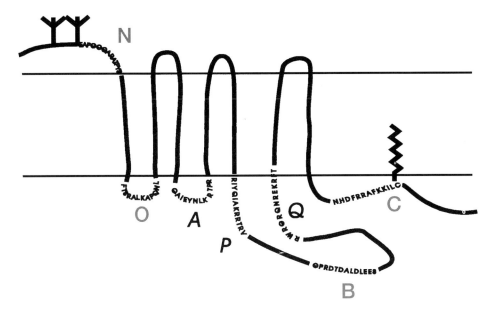

FIGURE 1 Location of α_{2A} adrenergic receptor peptides used to probe functional contact sites between receptor and G protein. (Modified from Dalman and Neubig, 1991).

GTPase assays. GTPase activity in G_o/G_i vesicles was measured using purified G_o/G_i from bovine brain reconstituted into asolectin vesicles as described (Wade *et al.*, 1996; Kim and Neubig, 1987).

Photoaffinity derivatives of Q peptide. The diazopyruvoyl (DAP) photoaffinity probe was coupled to the Q peptide (from the carboxy terminal end of the i3 loop of the α_{2a} adrenergic receptor) as described (Taylor *et al.*, 1994a, b). Photoincorporation of the DAP-Q into G_o α and β subunits was accomplished as described (Taylor *et al.*, 1996; Taylor *et al.*, 1994b). The sites of interaction were mapped by a combination of light tryptic digestion of non-denatured βγ and endoproteinase lys C fragmentation of labelled Gβ subunits in SDS-gel slices as described by Cleveland (Cleveland, 1983). Details of methods and results are described by Taylor *et al.* (1996).

Mass spectrometry and amino acid analysis. Mass spectrometry and amino acid analysis were performed by the University of Michigan Protein and Carbohydrate Structure Core Facility. Peptide masses and purity were determined by electrospray mass spectrometry using a Vestec single-quadrupole mass spectrometer with electrospray interface.

RESULTS AND DISCUSSION

Structure activity studies of Q peptide. Progressive decreases in GTPase stimulation by 100 μM peptide were observed as amino acids were removed from the amino-terminal end of the Q peptide (Wade *et al.*, 1996). This concentration gives maximal activity in the GTPase stimulation assay for intact Q peptide (Wade *et al.*, 1994). Stimulation was

TABLE 1

The EC_{50}'s of α_{2A} receptor peptides substituted in position 373 with various amino acids are compared with the degree of constitutive activation caused by the same substitutions in the receptor structure (data from Wade et al., 1996 and Ren et al., 1993).

Residue	EC_{50} (μM)	% basal inhibition
K	26	47
F	38	16
T	88	5
A	90	20
E	147	24

completely lost after deletion of 5 amino acids from the aminoterminal end of Q peptide (residues 361–365 in the receptor). It is notable that three of the five arginines in this peptide are in this portion of the peptide confirming the key importance of positive charges in compounds that activate G proteins *in vitro* (Higashijima *et al.*, 1990). Removal of amino acids from the carboxyl-terminal end of the Q peptide resulted in a more complex and interesting pattern of activity. Removal of the threonine in Q peptide (which corresponds to thr^{373} in the α_{2A} adrenergic receptor) actually enhanced peptide activity. Further truncations produced a decrease then increase in GTPase-stimulation activity. We were intrigued by the possibility that the low activity of the peptide with a carboxy-terminal threonine was related to the recent observation by Ren *et al.* (1993b) that replacement of residue 373 with other amino acids resulted in a "constitutively" activated α_2 adrenergic receptor. We prepared peptides with the same amino acids substituted for thr^{373}. All produced a similar maximum stimulation of 2.5–3.0 fold for GTPase stimulation but there were significant differences in potency (Wade *et al.*, 1996). Table 1 shows a summary of EC_{50}'s of 5 such peptides to activate the G_o/G_i GTPase compared to their degree of basal adenylyl cyclase inhibition reported by (Ren *et al.*, 1993a). Lysine produced a 3.4-fold increase in potency consistent with its substantial increase in basal adenylyl cyclase. Phenylalanine, which caused a small increase in basal adenylyl cyclase inhibition in the intact receptor showed an EC_{50} 2.3-fold lower than for the wild type peptide. In contrast, alanine which produced a bigger increase in basal activity than phenylalanine was identical in potency and most strikingly, glutamate, which was second best in its ability to induce constitutive activation of receptor caused a 50% decrease in potency of the peptide.

The absolute dependence of Q peptide activation of G protein on the region 361–365 implicates a key role for this positively charged region in the activation of G protein. It is possible that this is an effector region which directly activates the G protein. In contrast, the region predicted to be closer to the membrane is not required for G protein activation in the context of synthetic peptides. The fact that 3 residues from the BBXB motif (Okamoto and Nishimoto, 1992; Ikezu *et al.*, 1992) can be removed with retention of 60% of the GTPase stimulation further undermines the generality of this motif as a G protein activator. The highly variable effects of deletions in the membrane proximal portion of the i3c peptide are more consistent with a regulatory role for this portion of the peptide. The previously postulated conformational switch region (Lefkowitz *et al.*, 1993) which, when disrupted, results in constituitive activation of several G protein-coupled receptors is likely to be in this region. Our structure-activity data are consistent with such a role for the region of the i3 loop adjacent to TM6.

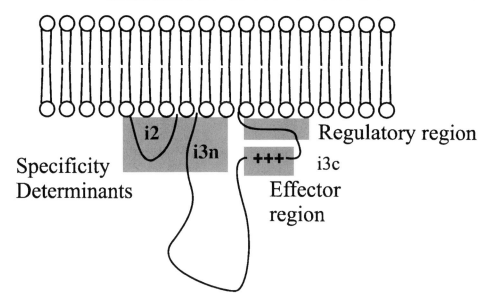

FIGURE 2 Proposed functions of intracellular loops of the α_{2A} adrenergic receptor based on mutagenesis and peptide structure-activity studies. The carboxyterminal end of the third intracellular loop appears to include both a regulatory domain near the membrane and a possible activator domain (or effector region) farther from the membrane which directly activates the G protein. The i2 and i3n regions may play a more important role in defining specificity for G proteins as mutations in these regions lead to loss of specificity.

The role of the i3c region of the m1 muscarinic receptor in the activation of phospholipase C has been studied in detail by a saturation mutagenesis approach (Burstein et al., 1995). Three amino acids (two alanines and a leucine) in the membrane-proximal i3c were never mutated in receptors with normal function. Of the remaining 21 residues, 5 were only mutated once in the receptors with normal function including the two arginines at positions 16 and 18 from the predicted membrane spanning domain (TM6). Two crucial arginines (RWR) in our Q peptide are at positions 15 and 17 from the predicted TM6 domain. Thus in both the m1 receptor mutagenesis and in our study, the BXBB structure 7 or 8 residues from TM6 appears to be less critical than the cluster of basic residues 10 amino acids farther out from the membrane.

This leaves uncertain the roles for the i2 and i3n regions of GPCRs. They have clearly been shown to be important for receptor-G protein coupling (Cheung et al., 1992; Pin et al., 1994; Wong and Ross, 1994; Wong et al., 1990). Much of the work implicating these regions is from chimer studies examining G protein coupling specificity. It is possible that they may be most important for specificity while the carboxyterminal end of i3 is a general G protein activator with a regulatory mechanism built in to prevent access to G protein in the basal state of the receptor. A scheme to incorporate these concepts is shown in Figure 2.

The regions of the α_{2a} AR and other 7TM receptors which interact with the G protein have been extensively studied. On the other end of this interaction, there is much information about the regions on α subunits which couple to receptors with the carboxy terminus playing a key role (Hamm et al., 1988; Masters et al., 1988; Conklin and Bourne, 1993). The amino terminus of Gα has been implicated as well (Taylor et al., 1994b;

Conklin and Bourne, 1993b). More recently, evidence supports a direct interaction of the βγ subunit with receptors. Binding of purified $β_1$-adrenergic receptors and rhodopsin with their respective G protein βγ subunits has been demonstrated (Phillips and Cerione, 1992; Phillips et al., 1992; Heithier et al., 1992). Kleuss et al. have shown that antisense probes directed against the $β_3$ subtype or the $γ_4$ subtype can block muscarinic receptor inhibition of calcium (Kleuss et al., 1992; Kleuss et al., 1993); while probes directed against $β_1$ or $γ_3$ block somatostatin receptor inhibition of calcium currents (Kleuss et al., 1992; Kleuss et al., 1993). Also, Kisselev and Gautam have shown that rhodopsin binds to a G protein containing $γ_1$ but not $γ_2$ or $γ_3$ subunits (Kisselev and Gautam, 1993). The isopreoid-modified carboxylterminus of γ subunit is important for coupling to rhodopsin (Kisselev et al., 1994). Thus we were very interested in identifying which subunit and more specifically which regions of the G protein βγ subunits were involved in receptor coupling.

We recently showed that the $α_{2A}$ AR-derived DAP-Q photoprobe specifically incorporated in the β subunit of G_o but did not label γ (Taylor et al., 1994b) thus directing our attention to the former subunit. The G protein β subunit is composed of 7 highly conserved WD-40 repeat regions characterized by a GH (Gly-His) followed by 23–41 core amino acids and a WD (Trp-Asp) (Neer et al., 1994). The function of these WD-40 repeats is not known, but it has been proposed that the repeats are important for protein-protein interactions (Neer et al., 1994).

Trypsin treatment of native βγ subunit results in cleavage of β at arg^{129} which generates a 14 kDa amino-terminal fragment and a 23 kDa carboxy-terminal fragment. Thomas et al., have shown trypsin treatment of native βγ subunits does not disrupt tertiary structures or the ability of the complex to associate functionally with the α subunit (Thomas et al., 1993). DAP-Q photo-labeled the 23 kDa carboxy-terminal trypsin fragment but not the 14 kDa fragment of the β subunit (Taylor et al., 1996). To further localize the major binding sites on the β subunit we labeled purified βγ with a radioiodinated DAP-Q digested the products with endoproteinase Lys C (20:1 protein:enzyme) according to Cleveland et al., (Cleveland, 1983). The majority of the radioactivity migrates as a ~6 kDa lys C fragment of the β subunit (Taylor et al., 1996). Labelling within residues 281 to 341 in $β_1$ or 302 to 341 in $β_2$ predicts labelled fragments of 5 or 7 kDa (Receptor sites, Figure 3). Since labelling within 281–301 (thin line) would generate a radio-labelled fragment of approximately 11.5 kDa from the $β_2$ subunit which is not observed, it is likely that both subunits are labelled within residues 302–341 (thick line in Figure 3). However, the ~6 kDa apparent molecular weight of the radio-labelled fragment is consistent also with cross-linking to $β_1$ in the region of 281–301 provided that cleavage at lys^{301} does not occur.

The only sites of β subunit labelling by DAP-Q are within the carboxy-terminal 24 kDa fragment of the β subunit and the majority of label corresponds to a site within residues 302–341 of either $β_1$ or $β_2$ or possibly 281–301 of $β_1$ (Figure 3). This region includes WD-40 repeat 7, the connecting loop to 6, and possibly repeat 6 itself (Figure 3). In the same study, labelling of the βγ subunit by the membrane-associated photoprobe [^{125}I] TID showed that substantial labelling of both the amino- and carboxy-terminal tryptic fragments of β while γ did not label.

In addition to the site on β described, we have also shown that DAP-Q labels the amino-terminus of $α_o$ (Taylor et al., 1994b). This same N-terminal region on α has been shown to bind to mastoparan (Higashijima and Ross, 1991) and to disrupt rhodopsin-transducin

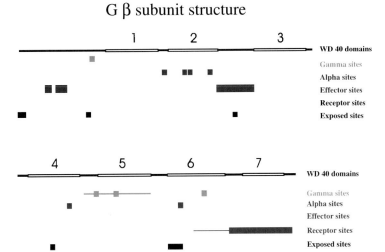

FIGURE 3 Proposed structural model and location of some functional sites in the Gβ sequence. WD40 domains are labelled "GH" to the "WD" and sites of interaction with γ, α, and effector are from published yeast genetic studies (see text). "Exposed sites" are locations of significant sized insertions in the yeast β compared to mammalian β plus the trypsin sensitive site in mammalian β (in the insert between WD's 2 and 3). Mapping of the receptor sites is described in (Taylor et al., 1996).

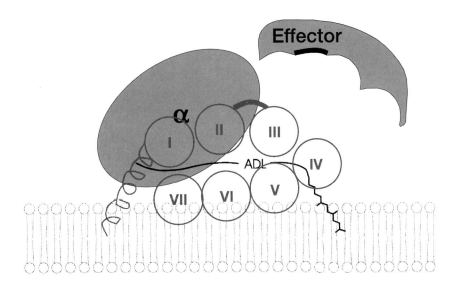

FIGURE 4 Predicted structural model of G protein membrane interactions. The Gβ subunit is shown as a series of circles representing the WD-40 domains plus an α helical amino terminal extension. The γ subunit is extended across the β subunit from the carboxyterminal prenyl groups shown interacting with the membrane to the aminoterminus in proximity to the aminoterminus of Gβ. The regions of Gβ and γ subunits which interact with each other, α subunit, effector, and with the membrane are predicted based on a review of the literature (see text). The α_2 adrenergic receptor i3c peptide labels the 6th or 7th WD-40 domain which are also shown in proximity to the membrane.

interactions (Hamm *et al.*, 1988). Although the regions on the β subunit that bind to α are not well defined, Neer and colleagues have shown that residues 204 and 271 in the carboxy-terminal 23 kDa fragment of $β_1$ can be chemically cross-linked to the α subunit (García-Higuera *et al.*, 1996). Studies of the interaction of Gβ and γ have implicated the amino-termini of both subunits and more extensive carboxyterminal regions of β (Garritsen and Simonds, 1994; Lee *et al.*, 1995; Katz and Simon, 1995). Most strikingly, there has been a large amount of detailed information from yeast genetic studies about regions of the G protein β and γ subunits which interact with each other (Whiteway *et al.*, 1992; García-Higuera *et al.*, 1996), with α subunit (Whiteway *et al.*, 1994), and with effector (Leberer *et al.*, 1992). proteins. By combining biochemical mapping results, chimers of mammalian β and γ, and data from yeast genetics (Figure 3) we propose a structural model of the βγ subunit and its disposition in the plasma membrane (Figure 4).

In this model, both the amino and carboxy terminal regions of β subunit interact with the membrane. The extensive contacts between β and γ are consistent with the γ subunit being extended across the β subunit. From Figure 3, it can be seen that α and the effector may have more limited contacts with β. The proximity of the α contact sites and effector contact sites from the yeast genetic work (i.e. WD domains II and III, respectively) are consistent with a steric model in which α and effector bind to neighboring regions of β in a mutually exclusive manner which accounts for the ability of free βγ subunit but not heterotrimer to activate effectors. Thus the heavy bar shown in the effector and in the linker between WD II and WD III represent complimentary contact sites during βγ activation of effector. Thus it may be the regions between the WD domains rather than the WD domains themselves which represent the crucial protein-protein contact sites required for function. The placement of the WD VI and VII near the membrane is consistent with labelling of this region by both the receptor-derived peptide and the hydrophobic photoaffinity label (Taylor *et al.*, 1996).

Thus we present two hypothetical structural models (one for receptor and one for G protein β subunits) which account for a range of data. The model of the $α_2$ adrenergic receptor intracellular loops is based on limited information because fine structure mapping of residues important for RG coupling has been difficult to obtain. In contrast, detailed genetic results from yeast as well as chimeric and biochemical studies with mammalian proteins have permitted a relatively constrained model of the G protein βγ subunit. In both cases, the models will be useful for developing and testing hypotheses, however, it will be necessary to further test the models or to obtain direct structural information by X-ray crystallography or NMR techniques.

ACKNOWLEDGMENTS

This work was supported by HLGM46417 (to RRN), the Michigan Arthritis Center (P60-AR20557), and pre-doctoral fellowships provided by the American Association of University Women and the University of Michigan Center for Protein Structure and Design (to JMT).

Note added in proof: Two X-ray crystal structures of G protein heterotrimer were published after submission of this manuscript (Wall *et al.*, Cell 83:1047–1058, 1995 and

Lambright *et al.*, Nature 379:311–319, 1996). The 3-D structure confirmed our prediction of the β subunit organization as a ring of 7 repeating units with an extended aminoterminus. In one difference from our model, the γ subunit drapes around the outside of the ring of WD40 repeats of the β subunit rather than across the middle as we suspected.

References

Brown, A.M. and Birnbaumer, L. (1990). Ionic channels and their regulation by G protein subunits. *Annu. Rev. Physiol.*, **52**, 197–213.

Burstein, E.S., Spalding, T.A., Hill-Eubanks, D. and Brann, M.R. (1995). Structure-function of muscarinic receptor coupling to G proteins. Random saturation mutagenesis identifies a critical determinant of receptor affinity for G proteins. *J. Biol. Chem.*, **270**, 3141–3146.

Cheung, A.H., Huang, R.R. and Strader, C.D. (1992). Involvement of specific hydrophobic, but not hydrophilic, amino acids in the third intracellular loop of the β-adrenergic receptor in the activation of G_s. *Mol. Pharmacol.*, **41**, 1061–1065.

Cleveland, D.W. (1983). Peptide mapping in one dimension by limited proteolysis of sodium dodecyl sulfate-solubilized proteins. *Methods. Enzymol.*, **96**, 222–229.

Conklin, B.R. and Bourne, H.R. (1993). Structural elements of Gα subunits that interact with Gβτ, receptors, and effectors. *Cell*, **73**, 631–641.

Dalman, H.M., Gerhardt, M.A. and Neubig, R.R. (1991). Differential effects of α_{2A}-adrenergic receptor peptides on G proteins. *FASEB J.*, **5**, A1594

Dalman, H.M. and Neubig, R.R. (1991). Two peptides from the α_{2A}-adrenergic receptor alter receptor G protein coupling by distinct mechanisms. *J. Biol. Chem.*, **266**, 11025–11029.

García-Higuera, I., Thomas, T.C., Yi, F. and Neer, E.J. (1996). Cross-linking residues in Gβ subunit to Gα. *J. Biol. Chem.*, **271**, 528–535.

Garritsen, A. and Simonds, W.F. (1994). Multiple domains of G protein β confer subunit specificity in βγ interaction. *J. Biol. Chem.*, **269**, 24418–24423.

Guyer, C.A., Horstman, D.A., Wilson, A.L., Clark, J.D., Cragoe, E.J.J. and Limbird, L.E. (1990). Cloning, sequencing and expression of the gene encoding the porcine α_2-adrenergic receptor. *J. Biol. Chem.*, **265**, 17307–17317.

Hamm, H.E., Deretic, D., Arendt, A., Hargrave, P.A., Koenig, B. and Hofmann, K.P. (1988). Site of G protein binding to rhodopsin mapped with synthetic peptides from the α subunit. *Science*, **241**, 832–835.

Hargrave, P.A., Hamm, H.E. and Hofmann, K.P. (1993). Interaction of Rhodopsin with the G-protein, transducin. *Bioessays*, **15**, 43–50.

Heithier, H., Frohlich, M., Dees, C., Baumann, M., Haring, M., Gierschik, P., Schiltz, E., Vaz, W.L., Hekman, M. and Helmreich, E.J. (1992). Subunit interactions of GTP-binding proteins. *Eur J Biochem.*, **204**, 1169–1181.

Hepler, J.R. and Gilman, A.G. (1992). G proteins. *Trends Biochem. Sci.*, **17**, 383–387.

Higashijima, T., Burnier, J. and Ross, E.M. (1990). Regulation of G_i and G_o by mastoparan, related amphiphilic peptides, and hydrophobic amines. Mechanism and structural determinants of activity. *J Biol. Chem.*, **265**, 14176–14186.

Higashijima, T. and Ross, E.M. (1991). Mapping of the mastoparan-binding site on G proteins. Cross-linking of [125I-Tyr3,Cys11]mastoparan to G_o. *J. Biol. Chem.*, **266**, 12655–12661.

Ikezu, T., Okamoto, T., Ogata, E. and Nishimoto, I. (1992). Amino acids 356–372 constitute a G_i-activator sequence of the α_2-adrenergic receptor and have a Phe substitute in the G protein-activator sequence motif. *FEBS Lett.*, **311**, 29–32.

Katz, A. and Simon, M.I. (1995). A segment of the C-terminal half of the G-protein β1 subunit specifies its interaction with the γ_1 subunit. *Proc. Natl. Acad. Sci. USA*, **92**, 1998–2002.

Kim, M.H. and Neubig, R.R. (1987). Membrane reconstitution of high-affinity α_2 adrenergic agonist binding with guanine nucleotide regulatory proteins. *Biochemistry*, **26**, 3664–3672.

Kisselev, O. and Gautam, N. (1993). Specific interaction with rhodopsin is dependent on the γ subunit type in a G protein. *J Biol. Chem.*, **268**, 24519–24522.

Kisselev, O.G., Ermolaeva, M.V. and Gautam, N. (1994). A farnesylated domain in the G protein γ subunit is a specific determinant of receptor coupling. *J. Biol. Chem.*, **269**, 21399–21402.

Kleuss, C., Scherubl, H., Hescheler, J., Schultz, G. and Wittig, B. (1992). Different β-subunits determine G-protein interactions with transmembrane receptors. *Nature*, **358**, 424–426.

Kleuss, C., Scherubl, H., Hescheler, J., Schultz, G. and Wittig, B. (1993). Selectivity in signal transduction determined by γ subunits of heterotrimeric G proteins. *Science*, **259**, 832–834.

Kobilka, B.K., Kobilka, T.S., Daniel, K., Regan, J.W., Caron, M.G. and Lefkowitz, R.J. (1988). Chimeric α_2, β_2-adrenergic receptors: delineation of domains involved in effector coupling and ligand binding specificity. *Science*, **240**, 1310–1316.

König, B., Arendt, A., McDowel, J.H., Kahlert, M., Hargrave, P.A. and Hofmann, K.P. (1989). Three cytoplasmic loops of rhodopsin interact with transducin. *Proc. Natl. Acad. Sci. USA*, **86**, 6878–6882.

Leberer, E., Dignard, D., Hougan, L., Thomas, D.Y. and Whiteway, M. (1992). Dominant-negative mutants of a yeast G-protein β subunit identify two functional regions involved in pheromone signalling. *EMBO J.*, **11**, 4805–4813.

Lee, C., Murakami, T. and Simonds, W.F. (1995). Identification of a discrete region of the G protein gamma subunit conferring selectivity in βγ complex formation. *J. Biol. Chem.*, **270**, 8779–8784.

Lefkowitz, R.J., Cotecchia, S., Samama, P. and Costa, T. (1993). Constitutive activity of receptors coupled to guanine nucleotide regulatory proteins. *Trends. Pharmacol. Sci.*, **14**, 303–307.

Malek, D., Münch, G. and Palm, D. (1993). Two sites in the third inner loop of the dopamine D_2 receptor are involved in functional G protein-mediated coupling to adenylate cyclase. *FEBS Lett.*, **325**, 215–219.

Masters, S.B., Sullivan, K.A., Miller, R.T., Beiderman, B., Lopez, N.G., Ramachandran, J. and Bourne, H.R. (1988). Carboxyl terminal domain of Gsa specifies coupling of receptors to stimulation of adenylyl cyclase. *Science*, **241**, 448–451.

Mukai, H., Munekata, E. and Higashijima, T. (1992). G protein antagonists: a novel hydrophobic peptide competes with receptor for G protein binding. *J. Biol. Chem.*, **267**, 16237–16243.

Munch, G., Dees, C., Hekman, M. and Palm, D. (1991). Multisite contacts involved in coupling of the β-adrenergic receptor with the stimulatory guanine-nucleotide binding regulatory protein. Structural and functional studies by β-receptor-site-specific peptides. *Eur. J. Biochem.*, **198**, 357–364.

Neer, E.J., Schmidt, C.J., Nambudripad, R. and Smith, T.F. (1994). The ancient regulatory-protein family of WD-repeat proteins. *Nature*, **371**, 297–300.

Neubig, R.R. and Dalman, H.M. (1991). Effect of α_{2a}-adrenergic receptor peptides on agonist binding to α_{2b}-adrenergic, muscarinic (M4) and opiate (delta) receptors in NG108-15 membranes. *FASEB J.*, **5**, A1594

Okamoto, T. and Nishimoto, I. (1992). Detection of G protein-activator regions in M_4 subtype muscarinic, cholinergic, and α_2-adrenergic receptors based upon characteristics in primary structure. *J. Biol. Chem.*, **267**, 8342–8346.

Phillips, W.J., Wong, S.C. and Cerione, R.A. (1992). Rhodopsin transducin interactions. II. Influence of the transducin-βτ subunit complex on the coupling of the transducin-α subunit to rhodopsin. *J. Biol. Chem.*, **267**, 17040–17046.

Phillips, W.J. and Cerione, R.A. (1992). Rhodopsin/Transducin interactions I. Characterization of the binding of the transducin-βτ subunit complex to rhodopsin using fluorescence spectroscopy. *J. Biol. Chem.*, **267**, 17032–17039.

Pin, J.P., Joly, C., Heinemann, S.F. and Bockaert, J. (1994). Domains involved in the specificity of G protein activation in phospholipase C-coupled metabotropic glutamate receptors. *EMBO J.*, **13**, 342–348.

Ren, Q., Kurose, H., Lefkowitz, R.J. and Cotecchia, S. (1993). Constitutively active mutants of the α_2-adrenergic receptor. *J. Biol. Chem.*, **268**, 16483–16487.

Strader, C.D., Sigal, I.S. and Dixon, R.A.F. (1989). Structural basis of β-adrenergic receptor function. *FASEB J.*, **3**, 1825–1832.

Taylor, J.M., Jacob-Mosier, G.G., Lawton, R.G. and Neubig, R.R. (1994a). Coupling an α_2-adrenergic receptor peptide to G-protein: A new photolabeling agent. *Peptides*, **15**, 829–834.

Taylor, J.M., Jacob-Mosier, G., Lawton, R.G., Remmers, A.E. and Neubig, R.R. (1994b). Binding of an α_2-adrenergic receptor peptide to Gβ and the N-terminus of Gα. *J Biol. Chem.*, **269**, 27618–27624.

Taylor, J.M., VanDort, M., Jacob-Mosier, G., Lawton, R.G. and Neubig, R.R. (1996). Receptor and membrane interaction sites on Gβ. A receptor-derived peptide binds to the carboxy-terminus. *J. Biol. Chem.*, **271**, 3336–3339.

Taylor, J.M. and Neubig, R.R. (1994). Peptides as probes for G protein signal transduction. *Cell. Signal.*, **6**, 841–849.

Thomas, T.C., Schmidt, C.J. and Neer, E.J. (1993). G protein α_o subunit: Mutation of conserved cysteines identifies a subunit contact surface and alters GDP affinity. *Proc. Natl. Acad. Sci. U.S.A.*, **90**, 10295–10299.

Varrault, A., Nguyen, D.L., McClue, S., Harris, B., Jouin, P. and Bockaert, J. (1994). 5-Hydroxytryptamine$_{1A}$ receptor synthetic peptides. *J. Biol. Chem.*, **269**, 16720–16725.

Voss, T., Wallner, E., Czernilofsky, A.P. and Freissmuth, M. (1993). Amphipathic α-helical structure does not predict the ability of receptor-derived synthetic peptides to interact with guanine nucleotide-binding regulatory proteins. *J. Biol. Chem.*, **268**, 4637–4642.

Wade, S.M., Dalman, H.M., Yang, S. and Neubig, R.R. (1994). Multisite interactions of receptors and G proteins. Enhanced potency of dimeric receptor peptides in modifying G protein function. *Mol. Pharmacol.*, **45**, 1191–1197.

Wade, S.M., Scribner, M.K. and Neubig, R.R. (1996). Structure-activity relations for receptor-derived peptides and their interactions with G proteins. *Mol. Pharmacol.*, in press.

Wagner, T., Oppi, C. and Tocchini-Valentini, G.P. (1995). Differential regulation of G protein α-subunit GTPase activity by peptdies derived from the third cytoplasmic loop of the α_2-adrenergic receptor. *FEBS Lett.*, **365**, 13–17.

Whiteway, M., Dignard, D. and Thomas, D.Y. (1992). Mutagenesis of Ste18, a putative Gγ subunit in the Saccharomyces cerevisiae pheromone response pathway. *Biochem. Cell Biol.,* **70**, 1230–1237.

Whiteway, M., Clark, K.L., Leberer, E., Dignard, D. and Thomas, D.Y. (1994). Genetic identification of residues involved in association of α and β G-protein subunits. *Mol. Cell Biol.,* **14**, 3223–3229.

Wong, S.K., Parker, E.M. and Ross, E.M. (1990). Chimeric muscarinic cholinergic: β-adrenergic receptors that activate G_s in response to muscarinic agonists. *J. Biol. Chem.,* **265**, 6219–6224.

Wong, S.K. and Ross, E.M. (1994). Chimeric muscarinic cholinergic: β-adrenergic receptors that are functionally promiscuous among G proteins. *J. Biol. Chem.,* **269**, 18968–18976.

DELINEATING LIGAND-SPECIFIC STRUCTURAL CHANGES IN ADRENERGIC RECEPTORS BY USE OF FLUORESCENCE SPECTROSCOPY

ULRIK GETHER,[§] SANSAN LIN[§] and BRIAN K. KOBILKA[‡§]

[§]*Howard Hughes Medical Institute,* [‡]*Division of Cardiovascular Medicine, Stanford University Medical School, Stanford, California 94305*

The conformational state of a G protein coupled receptor has previously only been assessed by indirect methods, such as the effect of receptor conformation on G protein GTPase activity or on the activity of the effector enzymes. The purpose of the present studies has been to develop new techniques that would enable us to directly monitor structural changes in a G protein coupled receptor in a pure, isolated system. We have taken advantage of the sensitivity of many fluorescent molecules to the polarity of their molecular environment. Fluorescent labels incorporated into proteins can thus be used as sensitive molecular reporters of conformational changes and of protein-protein interactions that cause changes in polarity of the environment surrounding the probe. The human β_2 adrenergic receptor was expressed in SF-9 insect cells, purified and subsequently covalently labeled with the cysteine-reactive, fluorescent probe IANBD (N,N'-dimethyl-N(iodoacetyl)-N'-(7-nitrobenz-2-oxa-1,3-diazol-4-yl) ethylenedi-amine). It was found that adrenergic agonists caused a dose-dependent and antagonist-reversible decrease in the fluorescence from IANBD labeled β_2 receptor. The change in fluorescence from the labeled β_2 receptor displayed a linear correlation with the biological efficacy of the individual compound. This suggests that the agonist-mediated decrease in fluorescence from IANBD labeled β_2 receptor describes the same conformational change as that involved in receptor activation and G protein coupling. In contrast to agonists, negative antagonists induced a small but significant increase in base-line fluorescence, which correlated with the negative intrinsic activity observed in biological assays. These data support the concept that antagonists by themselves can alter receptor structure. In conclusion, our data propose the possible existence of several ligand-specific conformational states of G protein coupled receptors. Furthermore, the data demonstrate the potential of fluorescence spectroscopy as a tool for further delineating the molecular mechanisms of drug action at G protein coupled receptors. Using a series of mutant β_2 receptors with a limited number of reactive cysteines we are currently attempting to map the fluorescence changes to specific subdomains of the receptor. In this way it should be possible to more precisely define the molecular mechanism of transmembrane signal transduction in G protein coupled receptors.

KEY WORDS: G protein coupled receptors, β_2 adrenergic receptor, conformational changes, fluorescence, inverse agonists, receptor activation

INTRODUCTION

The adrenergic receptors belong to the G protein coupled receptor family which now include several hundreds different receptors characterized by a remarkable diversity among their endogenous ligands (Hein and Kobilka, 1995; Savarese and Fraser, 1992). The receptors are similar to bacteriorhodopsin believed to share a common topology with

---

Correspondence to: Ulrik Gether, MD, Howard Hughes Medical Institute, B159 Beckman Center, Stanford University Medical School, Stanford CA 94305. Tel. 415 725 7754 Fax. 415 498 5092

seven α-helical, transmembrane segments. The helical arrangement and actual three-dimensional structure of the receptors, however, are still unknown (Hein and Kobilka, 1995; Savarese and Fraser, 1992; Schwartz, 1994). The current knowledge about structure-function relationships in the receptor family is so far almost exclusively based on mutagenesis studies performed over the last decade since the β_2 adrenergic receptor was cloned as the first G protein coupled receptor (Hein and Kobilka, 1995; Savarese and Fraser, 1992; Schwartz, 1994). These mutagenesis studies in the β_2 receptor as well as in many other G protein coupled receptors have allowed assignment of distinct receptor functions, like for example ligand binding and G protein coupling, to specific receptor domains (Hein and Kobilka, 1995; Savarese and Fraser, 1992; Schwartz, 1994). Nevertheless, the critical molecular events and related conformational changes providing the crucial link between agonist binding to the receptor and activation of the associated G protein are still unresolved.

The classical model for receptor action predicts that receptors are simple bimodal switches with an unliganded "off" configuration and an agonist dependent "on" configuration (Stephenson, 1956). The view has been challenged by the finding that several G protein coupled receptors in the absence of agonist have a significant level of basal activity (Costa and Herz, 1989; Schutz and Freissmuth, 1992; Samama et al., 1993; Lefkowitz et al., 1993; Samama et al., 1994; Chidiac et al., 1994; Barker et al., 1994; Bond et al., 1995), which interestingly can be enhanced by selective mutations in the third intracellular loop (Samama et al., 1993; Shenker et al., 1993; Barker et al., 1994; Parma et al., 1993). These observations have been explained by a two-state model where the receptors are proposed to exist in a dynamic equilibrium between an inactive state (R) and an active state (R*) (R $\Leftrightarrow$ R*) (Costa and Herz, 1989; Schutz and Freissmuth, 1992; Samama et al., 1993; Lefkowitz et al., 1993; Samama et al., 1994; Kenakin, 1994; Bond et al., 1995); and where the biological response to a given ligand is governed by its intrinsic ability to change the overall equilibrium between the two states (Costa and Herz, 1989; Schutz and Freissmuth, 1992; Samama et al., 1993; Lefkowitz et al., 1993; Samama et al., 1994; Kenakin, 1994; Bond et al., 1995). An alternative to the two-state model would be a multi-state-model in which the functional consequence of binding different classes of ligands is explained by the existence of several ligand-induced conformational states, where agonists, partial agonists, neutral antagonists and negative antagonists (also referred to as inverse agonists) may stabilize distinct receptor conformations (Gether et al., 1993; Schambye et al., 1994; Chidiac et al., 1994; Elling et al., 1995; Chidiac, 1995; Gether et al., 1995b).

To date, the conformational state of G protein coupled receptors has only been assessed by indirect methods, such as the effect of receptor conformation on G protein GTPase activity or on the activity of the effector enzymes. The purpose of our current studies has been to develop new techniques, which enable us to directly monitor structural changes in a G protein coupled receptor in an isolated pure system. Our approach takes advantage of the sensitivity of many fluorescent molecules to the polarity of their molecular environment (Cerione, 1994; Phillips and Cerione, 1991; Dunn and Raftery, 1993; Gettins et al., 1993). Fluorescent labels incorporated into proteins can therefore often be used as sensitive reporters of conformational changes and of protein-protein interactions that cause changes in polarity of the environment surrounding the probe (Cerione, 1994; Phillips and Cerione, 1991; Dunn and Raftery, 1993; Gettins et al., 1993).

FLUORESCENT LABELING OF THE β_2 ADRENERGIC RECEPTOR

To obtain sufficient amount of pure receptor protein needed for performing the fluorescence spectroscopy analysis, we expressed the human β_2 adrenergic receptor in SF-9 insect cells. The receptor was tagged at the aminoterminus with the cleaveable influenza-hemagglutinin signal-sequence followed by the M1 FLAG antibody epitope (Kobilka, 1995; Gether et al., 1995a) and at the carboxyterminus with six histidines. A purification procedure was then established with Ni-column chromatography as the initial step followed by purification on an M1 anti-FLAG-antibody column and finally alprenolol-affinity chromatography (Kobilka, 1995; Gether et al., 1995a).

The purified receptor was labeled with the cysteine-specific and environmentally sensitive fluorescent probe IANBD [N,N'-dimethyl-N (iodoacetyl)-N'-(7-nitrobenz-2-oxa-1,3-diazol-4-yl) ethylenediamine]. The fluorescence of IANBD is highly dependent on the polarity of the solvent and is more than ten fold stronger in n-butanol and n-hexane than in aqueous buffer (Figure 1A). There is a parallel blue-shift in the emission maximum from 540 nm in aqueous buffer to 530 nm in n-butanol and 510 nm in n-hexane (Figure 1A). The β_2 adrenergic receptor labeled with IANBD at a stoichiometry of about 1.2 moles of IANBD per mole of receptor demonstrated a strong fluorescence signal with an emission maximum at 523 nm (Figure 1B). The blue-shift in emission maximum, as compared to cysteine-reacted IANBD in aqueous buffer, indicated that the modified cysteine(s) are located in an environment that, on the average, is of lower polarity than n-butanol but higher than n-hexane (Figure 1B). This would likely involve labeling of one or more of the five cysteine residues that are located in the transmembrane, hydrophobic core of the receptor. The covalent modification of the receptor was confirmed by SDS-polyacrylamide gel electrophoresis of the labeled receptor, and the specificity of the labeling was verified by blocking the incorporation of IANBD with the cysteine-specific, non-fluorescent reagents, iodoacetamide and N-ethylmaleimide (Figure 1B, *insert*). The fluorescent labeling did not perturb the pharmacological properties of the receptor as both agonist and antagonist binding was unchanged (Gether et al., 1995a).

AGONIST-MEDIATED CHANGES IN FLUORESCENCE FROM IANBD LABELED β_2 RECEPTOR

Binding of the full agonist, isoproterenol, to IANBD labeled β_2 receptor caused a decrease in fluorescence intensity without detectable change in the wavelength at which maximal emission occurred (Gether et al., 1995a). The decrease was stereospecific as demonstrated by comparing the effect of the active (−)-isomer of isoproterenol with the less (+)-isomer (Gether et al., 1995a). To investigate the kinetics of this agonist mediated change, the fluorescence intensity was measured as a function of time (Figure 2). Prior to adding ligand we observed a slight but constant decline in baseline fluorescence (Figure 2). This loss of fluorescence over time is likely due to factors such as bleaching and hydrolysis of the probe during the experiments. It is unlikely that this decline in fluorescence is due to denaturation of the protein, since a similar loss of fluorescence also was observed with labeled receptor that was intentionally denatured in guanidinium chloride (Gether et al., 1995a).

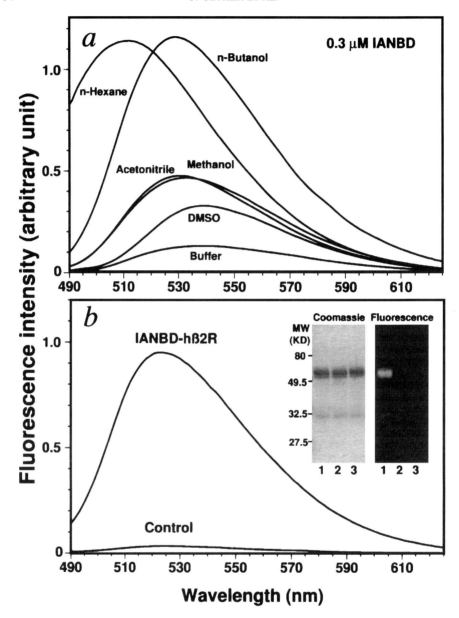

FIGURE 1 Fluorescence properties of IANBD and IANBD-labeled β_2 receptor. A, Emission spectra of cysteine-reacted IANBD (0.3 μM) in solvents of different polarity. Excitation was set at 481 nm. B, Emission spectrum of IANBD labeled β_2 receptor (0.15 μM receptor, 1.2 mol IANBD per mol receptor). Control is emission spectrum of 0.15 μM β_2 receptor 'labeled' with IANBD prebound to free cysteine instead of free IANBD to assess possible non-specific attachment of the probe to the receptor during labeling. *Insert;* 10% SDS-polyacrylamide gel electrophoresis of IANBD labeled β_2 receptor. Lane 1, 150 pmol IANBD labeled β_2 receptor; lane 2 and 3, 150 pmol β_2 receptor preincubated before exposure to IANBD with iodoacetamide (lane 2) and N-ethylmaleimide (lane 3). *Left panel* of insert, Commassie Blue staining of gel; *right panel* of insert, gel photographed under UV light. The weak band with an apparent molecular weight of 32.5 kD is a degradation product of the receptor. (Reproduced with permission from Gether *et al.,* 1995a, *J. Biol. Chem.*, **270**, 28268–28275.)

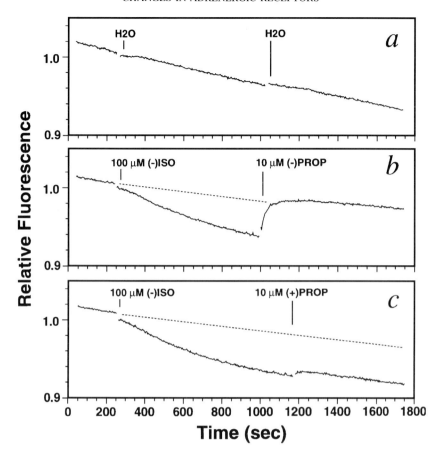

FIGURE 2 Reversibility of isoproterenol-induced decrease in fluorescence from IANBD labeled β_2 receptor. A, Control addition of water (H$_2$O). B and C, Reversal of the response to isoproterenol (ISO) by the active (−)-isomer of the antagonist propranolol, (−)PROP (B), but not by the less active (+)-isomer, (+)PROP (C). *Dotted lines* indicate extrapolated baseline. Excitation was 481 nm and emission measured at 523 nm. Fluorescence in all the individual traces shown was normalized to the fluorescence observed immediately after addition of ligand. All traces shown are representative of at least three identical experiments. (Reproduced with permission from Gether *et al.*, 1995a, *J. Biol. Chem.*, **270**, 28268–28275.)

The time course analysis showed that the response to isoproterenol reached a maximum amplitude below the extrapolated baseline after 10 to 15 min (Figure 2). The response could be readily reversed by the active (−)-isomer of the antagonist propranolol but not by the less active (+)-isomer. The response to isoproterenol was similarly reversed by several other antagonists, including alprenolol, ICI 118,551, pindolol and dichloroisoproterenol (data not shown). The response was also dose-dependent and the relative change in fluorescence at t = 15 min following addition of isoproterenol could be plotted against the logarithm of the isoproterenol concentration showing the best fit to a one-site sigmoid curve with an EC$_{50}$ of 19 µM (Gether *et al.*, 1995a). The EC$_{50}$ value of 19 µM for isoproterenol is higher than the binding constants observed using conventional radioligand binding techniques (~1 µM) (Gether *et al.*, 1995a). One explanation for this apparent

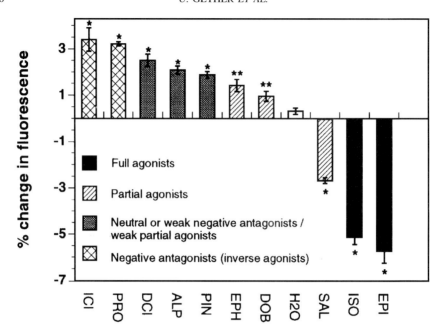

FIGURE 3 Effect of different adrenergic ligands on fluorescence from IANBD labeled β_2 receptor. The percent change (mean ± SE) in fluorescence was calculated as the change in fluorescence relative to the extrapolated baseline at t = 15 min after addition of ligand. The ligands and concentrations used were (number of experiments in parenthesis); ICI, 10^{-5} M ICI 118,551 (n = 4); PRO, 10^{-5} M (–)propranolol (n = 3); DCI, 10^{-4} M dichloroisoproterenol (n = 3); ALP, 10^{-4} M (–)alprenolol (n = 3); PIN, 10^{-5} M pindolol (n = 3); EPH, 10^{-3} M ephedrin (n = 3); DOB, 10^{-3} M dobutamine (n = 3); H$_2$O, water (n = 5), SAL, 10^{-3} M salbutamol (n = 3); ISO, 10^{-3} M isoproterenol (n = 3) and EPI, 10^{-3} M epinephrine (n = 3). Responses significantly different from water are indicated by *, p < 0.0005 (unpaired t test). The ligand concentrations used were chosen to ensure full saturation of the receptor with all ligands. This was confirmed by the ability of all compounds (at the concentration used) to fully displace ^{3}H-dihydroalprenolol from the receptor in a binding assay (data not shown). All of the tested compounds had an absorbance of less than 0.01 AU at 481 nm and 523 nm in the concentrations used excluding any 'inner filter' effect.

discrepancy could be technical differences in the methods by which the binding constants were obtained [the fluorescence studies were done with more than 100 fold higher receptor concentrations (100 nM receptor) for 15 min.]. Although apparent maximal change in fluorescence was observed already after 15 minutes, full equilibrium may not have been reached. Unfortunately, the fluorescent change upon ligand binding cannot be reliably determined after one hour of incubation due to magnification of baseline differences over a longer period of time. Another potential explanation for the difference in apparent K_D values would be incomplete labeling of the cysteine in the β_2 receptor, which causes the agonist induced response. The fraction of labeled receptors could exhibit a lowered agonist affinity in contrast to the fraction of receptors unlabeled at this site. However, this explanation is not supported by our radioligand binding data, which only detected a single agonist and antagonist affinity site and by the fact that we do not observe a change in the total number of binding sites following labeling with IANBD (Gether *et al.*, 1995a).

A series of adrenergic agonists having different biological efficacy were tested for their ability to affect fluorescence from IANBD labeled β_2 receptor (Figure 3). The changes in fluorescence were then compared with the biological efficacy of agonists in an adenylate cyclase assay using membranes from SF-9 cells expressing the β_2 adrenergic receptor (Gether et al., 1995a). The full agonists, epinephrine and isoproterenol, which produced the largest increase in adenylate cyclase activity, also caused the largest decrease in fluorescence (5–6%); whereas the relatively strong partial agonist salbutamol decreased fluorescence about 2.6% in agreement with a lower intrinsic activity in the adenylyl cyclase assay. The weaker partial agonists, dobutamine and ephedrine, did not cause a decrease in fluorescence relative to the extrapolated baseline but rather produced a slight increase (Figure 3). Interestingly, plotting percent change in baseline fluorescence induced by this series of adrenergic agonists versus percent change in basal adenylate cyclase activity revealed a clear linear correlation. This strongly suggests that the agonist-mediated change in fluorescence describes the conformational change in the receptor which is involved in receptor activation and G protein coupling.

The agonist induced change most likely represents changes in the polarity of the environment surrounding one of more labeled cysteines in the hydrophobic core of the protein. One possible mechanism could be that the agonist induces movement of a transmembrane segment perpendicular to the plane of the lipid bilayer which could result in the exposure of the fluorophore to the solvent. This would cause quenching of the fluorescence and thus a decrease in the net fluorescence from the labeled receptor. Alternatively, it could be imagined that the agonist could induce a rotation of the membrane spanning domain resulting in movement of the fluorophore from pointing toward the lipid bilayer to pointing into the core of the protein, which is predicted to be more polar. According to the first model, the fluorophore should be more accessible to an aqueous quencher like potassium iodide following agonist binding. However, we have observed that the agonist isoproterenol caused a slight decrease in quenching of the fluorescence from the IANBD labeled receptor (Gether et al., 1995a). This would favor the latter model in which the fluorophore is moved to a more hydrophilic pocket in the core of the protein following agonist binding. In this model aqueous quenchers might be expected to have a more limited access to the fluorophore. It should be noted that at this point we can only speculate on the molecular mechanism behind the ligand induced changes in fluorescence. With a stoichiometry of labeling at 1.2 mole IANBD per mole of receptor at least two sites are being labeled in the receptor. Thus, the isoproterenol-mediated decrease in fluorescence may involve a different labeled cysteine than that responsible for the isoproterenol induced effect on potassium iodide quenching.

ANTAGONIST-MEDIATED CHANGES IN FLUORESCENCE FROM IANBD LABELED β_2 RECEPTOR

We have also investigated whether antagonist by themselves were able to change the fluorescence from the IANBD labeled β_2 receptor. Propranolol and ICI 118,551, which both have been described as so-called negative antagonists (Chidiac et al., 1994; Bond et al., 1995; Samama et al., 1994), were found to cause a relative *increase* in fluorescence intensity from IANBD-labeled β_2 receptor followed by apparent stabilization of the rate at which fluorescence decreased over time (Figure 3 and (Gether et al., 1995a)). The response to propranolol was stereo-selective since it was not elicited by the less active

(+)-propranolol, indicating that the change must involve binding of propranolol to the receptor (Gether *et al.*, 1995a). Alprenolol, pindolol and dichloroproterenol, which in biological assays have been shown to exhibit either weak partial agonism, neutral antagonism or negative antagonism (Jasper and Insel, 1992; Chidiac *et al.*, 1994; Samama *et al.*, 1994; Samama *et al.*, 1993), induced a smaller but nevertheless reproducible increase in fluorescence (Figure 3).

It has previously been suggested, but never structurally verified, that antagonists may stabilize a conformation of the receptor that is distinct from unliganded receptor (Gether *et al.*, 1993; Schambye *et al.*, 1994; Chidiac *et al.*, 1994; Elling *et al.*, 1995; Chidiac, 1995; Gether *et al.*, 1995b). For example, this has been proposed to explain the unexpected observation that nonpeptide antagonists of G protein coupled peptide receptors can act as competitive antagonists for agonist peptides without sharing apparent binding sites in the receptor (Gether *et al.*, 1993; Schambye *et al.*, 1994; Elling *et al.*, 1995; Gether *et al.*, 1995b). In other words, agonists and antagonists may be able to mutually exclude each others binding to the receptor by stabilizing different receptor conformations (Gether *et al.*, 1993; Schambye *et al.*, 1994; Elling *et al.*, 1995; Gether *et al.*, 1995b). It was therefore intriguing to observe that this series of adrenergic antagonists produced small but very reproducible increases in baseline fluorescence from the IANBD labeled β_2 receptor suggesting that antagonists by themselves can alter receptor structure (Figure 3). Except in the case of dichloroisoproterenol, the changes showed a correlation with the negative intrinsic activity of the tested compounds. Thus, propranolol and ICI 188,551, which exhibited the strongest negative intrinsic activity, caused the largest increase in baseline fluorescence, whereas alprenolol and pindolol with a smaller negative intrinsic activity also caused a smaller increase in fluorescence. With respect to dichloroisoproterenol, it was surprising to find that it increased fluorescence to a similar extent as the negative antagonists and at the same time was able to activate the receptor. It is tempting to speculate that this increase may reflect a conformational change that is distinct from the change induced by the full agonists and the negative antagonists, but still is able to activate G_s. However, given the small size of the antagonist responses it is difficult to assess their exact molecular significance at the present stage.

MODEL FOR ACTIVATION OF G PROTEIN COUPLED RECEPTORS: TWO-STATE VERSUS MULTI-STATE?

The present fluorescence studies represent the first attempts to directly assess the conformational state of a G protein coupled receptor in response to binding of different classes of ligands. As a molecular reporter we have used the cysteine-selective and environmentally sensitive, fluorescent probe, IANBD, which can be covalently incorporated into the purified, human β_2 adrenergic receptor without perturbing the pharmacological properties of the receptor. The data provide structural evidence in a pure system for the existence of ligand-specific conformational states of a G protein coupled receptor. It is of interest to compare our data with prevailing models for activation of G protein coupled receptors, since the conformational state of a G protein coupled receptor previously only has been assessed by indirect methods. It is, however, very important to note that without knowing the exact molecular mechanism behind the fluorescent changes we must be very cautious when interpreting the data according to different models of activation. The following considerations should therefore be considered a starting point for a discussion rather than anywhere near a definitive conclusion.

The two-state model for activation of G protein coupled receptors predicts that the receptors exist in a dynamic equilibrium between two states, an inactive (R) and active conformation (R*). Agonists are believed to preferentially bind to and stabilize R* and thereby strongly shift the equilibrium to the right, whereas partial agonists are considered less effective in stabilizing R* (Costa and Herz, 1989; Schutz and Freissmuth, 1992; Samama et al., 1993; Lefkowitz et al., 1993; Samama et al., 1994; Kenakin, 1994; Bond et al., 1995). Conversely, negative antagonists are believed to preferentially bind to and stabilize R and shift the equilibrium to the left, whereas neutral antagonists block agonist binding but do not alter the equilibrium between R and R* (Costa and Herz, 1989; Schutz and Freissmuth, 1992; Samama et al., 1993; Lefkowitz et al., 1993; Samama et al., 1994; Kenakin, 1994; Bond et al., 1995). According to this model the fluorescence properties of the unoccupied IANBD labeled receptor would be expected to represent the average fluorescence properties of the population of β_2 receptors in the inactive state (R) and the active state (R*). The fluorescent properties of the R state alone should be observed in the presence of a negative antagonist, while those of the R* state alone should be observed in the presence of a full agonist (Costa and Herz, 1989; Schutz and Freissmuth, 1992; Samama et al., 1993; Lefkowitz et al., 1993; Samama et al., 1994; Kenakin, 1994; Bond et al., 1995). Using our fluorescence assay we first compared a series of adrenergic agonists, which exhibited distinct efficacies in a biological assay (Gether et al., 1995a). These data can be interpreted in agreement with the two-state model. Hence, the different functional efficacies may result from differences in the ability of the different agonists to pull the equilibrium towards R* and thus decrease the fluorescence signal (Gether et al., 1995a). The antagonist data are less obvious to explain within the two-state model unless the proportion of R* relative to R is significantly increased for the purified receptor. Thus, unoccupied β_2 receptors are predicted to exist primarily in the inactive R state (Costa and Herz, 1989; Schutz and Freissmuth, 1992; Samama et al., 1993; Lefkowitz et al., 1993; Samama et al., 1994; Kenakin, 1994; Bond et al., 1995). Therefore, it would not be expected to observe a significant change in the fluorescent properties of purified β_2 receptor on binding to neutral antagonists like pindolol and alprenolol, which are predicted not to change the distribution of R and R*, or negative antagonists, like ICI 118,551 and propranolol, which would be predicted to change the pool of receptor in the inactive conformation by only a few percent. This is also consistent with previous proteolysis studies on membrane bound receptor which show that agonists and antagonists are equivalent in protecting the β_2 receptor from proteolysis (Kobilka, 1990). This would not be expected if antagonists only changed the conformation of a small percentage of the receptor population.

Alternatively, the data could also support a multi-state model in which the biological efficacy of an agonist may be a consequence of the magnitude of conformational change that it induces in the receptor, rather than just affecting the equilibrium between only two states. This model is illustrated in Figure 4. In contrast to the two-state model, this model does not require the existence of two distinct conformations of the unliganded receptor (R and R*). According to this model the unliganded receptor has a specific rather low affinity for the G protein, but high enough to produce agonist independent activity if

a) Two-state-model:

b) Multi-state-model:

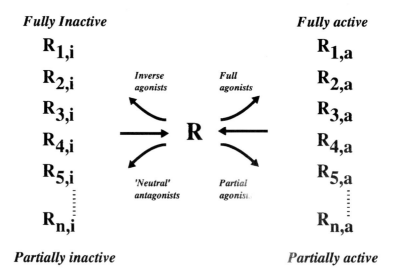

FIGURE 4 Models for activation of G protein coupled receptors. A, The two-state model for activation of G protein coupled receptors predicts that the receptor exist in a dynamic equilibrium between two states, an inactive (R) and active conformation (R*). Agonists are believed to preferentially bind to and stabilize R* and thereby strongly shift the equilibrium to the right, whereas partial agonists are considered less effective in stabilizing R*. Conversely, negative antagonists are believed to preferentially bind to and stabilize R and shift the equilibrium to the left, whereas neutral antagonists block agonist binding but do not alter the equilibrium between R and R* (Costa and Herz, 1989; Schutz and Freissmuth, 1992; Samama et al., 1993; Lefkowitz et al., 1993; Samama et al., 1994; Kenakin, 1994; Bond et al., 1995). B, a multi-state model. The multi-state model does not require the existence of two distinct conformations of the unliganded receptor (R and R*). According to this model the unliganded receptor has a specific rather low affinity for the G protein, but high enough to produce agonist independent activity if expressed at a sufficiently high level. Agonist binding produces a conformational change that increases the affinity of the receptor for the G protein whereas a negative antagonists cause a conformational change that decreases the affinity of the receptor for the G protein. The conformational change promoted by partial agonists results in an intermediate affinity for the G protein whereas 'neutral' antagonists either cause no or only a very minor change in receptor conformation. Thus, agonists and antagonists can be considered positive and negative allosteric modulators, respectively, of the affinity of the receptor for the G protein and thereby of the affinity of the G protein for GDP versus GTP.

expressed at a sufficiently high level. A constitutively activated receptor (Samama *et al.*, 1993; Shenker *et al.*, 1993; Barker *et al.*, 1994; Parma *et al.*, 1993) is in this model a receptor which intrinsic G protein affinity is increased. Agonist binding produces a conformational change that increases the affinity of the receptor for the G protein whereas a negative antagonists cause a conformational change that decreases the affinity of the receptor for the G protein. The conformational change promoted by partial agonists results in an intermediate affinity for the G protein whereas 'neutral' antagonists either cause no or only a very minor change in receptor conformation. Thus, agonists and antagonists can be considered positive and negative allosteric modulators, respectively, of the affinity of the receptor for the G protein and thereby of the affinity of the G protein for GDP versus GTP. The model does not make any assumptions whether the receptor spontaneously adopts the different conformational states, which then are stabilized by the ligands, or whether the distinct conformational states are promoted by an induced fit between the ligands and the receptor.

References

Barker, E.L., Westphal, R.S., Schmidt, D. and Sander-Bush, E. (1994). Constitutively active 5-hydroxytryptamine-2C receptors reveal novel inverse agonist activity of receptor ligands. *J. Biol. Chem.,* **269**, 11687–11690.
Bond, R.A., Leff, P., Johnson, T.D., Milano, C.A., Rockman, H.A., McMinn, T.R., Apparsundaram, S., Hyek, M.F., Kenakin, T.P., Allen, L.F. and Lefkowitz, R.J. (1995). Physiological effects of inverse agonists in transgenic mice with myocardial expression of the β_2-adrenergic receptor. *Nature,* **374**, 272–276.
Cerione, R.A. (1994). Fluorescence assays for G-protein interactions. *Method. Enzymol.,* **237**, 409–423.
Chidiac, P. (1995). Receptor state and ligand efficacy. *Trends Pharmacol. Sci.,* **16**, 83–84.
Chidiac, P., Hebert, T.E., Valiquette, M., Dennis, M. and Bouvier, M. (1994). Inverse agonist activity of β-adrenergic antagonists. *Mol. Pharmacol.,* **45**, 490–499.
Costa, T. and Herz, A. (1989). Antagonists with negative intrinsic activity at δ-opioid receptors coupled to GTP-binding proteins. *Proc. Natl. Acad. Sci. USA,* **86**, 7321–7325.
Dunn, S.M.J. and Raftery, M.A. (1993). Cholinergic binding sites on the pentameric binding site on the acetylcholine receptor of Torpedo Californica. *Biochemistry,* **32**, 8608–8615.
Elling, C.E., Nielsen, S.M. and Schwartz, T.W. (1995). Conversion of antagonist-binding site to metal-ion site in the tachykinin NK-1 receptor. *Nature,* **374**, 74–77.
Gether, U., Johansen, T.E., Snider, R.M., Lowe, J.A., Nakanishi, S. and Schwartz, T.W. (1993). Different binding epitopes on the NK1 receptor for substance-P and a non-peptide antagonist. *Nature,* **362**, 345–348.
Gether, U., Lin, S. and Kobilka, B.K. (1995a). Fluorescent labeling of purified β_2-adrenergic receptor: Evidence for ligand-specific conformational changes. *J. Biol. Chem.,* **270**, 28268–28275.
Gether, U., Lowe III, J.A. and Schwartz, T.W. (1995b). Tachykinin non-peptide antagonists: binding domain and molecular mode of action. *Biochem. Soc. Trans.,* **23**, 96–102.
Gettins, P.G.W., Fan, B., Crews, B.C. and Turko, I.V. (1993). Transmission of conformational change from the heparin binding site to the reactive center of antithrombin. *Biochemistry,* **32**, 8385–8389.
Hein, L. and Kobilka, B.K. (1995). Adrenergic receptor signal transduction and regulation. *Neuropharmacology,* **34**, 357–366.
Jasper, J.R. and Insel, P.A. (1992). Evolving concepts of partial agonism: The β-adrenergic receptor as a paradigm. *Biochem. Pharmacol.,* **43**, 119–130.
Kenakin, T. (1994). On the definition of efficacy. *Trends Pharmacol. Sci.,* **15**, 408–409.
Kobilka, B.K. (1990). The role of cytosolic and membrane factors in processing of the human β_2 adrenergic receptor following translocation and glycosylation in a cell-free system. *J. Biol. Chem.,* **265**, 7610–7618.
Kobilka, B.K. (1995). Amino and carboxyterminal modifications to facilitate production and purification of a G protein coupled receptor. *Anal. Biochem.,* **231**, 269–271.
Lefkowitz, R.J., Cotecchia, S., Samama, P. and Costa, T. (1993). Constitutive activity of receptors coupled to guanine nucleotide regulatory proteins. *Trends Pharmacol. Sci.,* **14**, 303–307.
Parma, J., Duprez, L., Van Sande, J., Cochaux, P., Gervy, C., Mockel, J., Dumont, J. and Vassart, G. (1993). Somatic mutations in the thyrotropin receptor gene cause hyperfunctioning thyroid adenomas. *Nature,* **365**, 649–651.
Phillips, W.J. and Cerione, R.A. (1991). Labeling of the β-γ subunit complex of transducin with an environmentally sensitive cysteine reagent. Use of fluorescence spectroscopy to monitor transducin subunit interactions. *J. Biol. Chem.,* **266**, 11017–11024.

Samama, P., Cotecchia, S., Costa, T. and Lefkowitz, R.J. (1993). A mutation-induced activated state of the β_2-adrenergic receptor: Extending the ternary complex model. *J. Biol. Chem.,* **268**, 4625–4636.

Samama, P., Pei, G., Costa, T., Cotecchia, S. and Lefkowitz, R.J. (1994). Negative antagonists promote an inactive conformation of the β_2-adrenergic receptor. *Mol. Pharmacol.,* **45**, 390–394.

Savarese, T.M. and Fraser, C.M. (1992). *In vitro* mutagenesis and the search for structure-function relationships among G protein-coupled receptors. *Biochem. J.,* **283**, 1–19.

Schambye, H.T., Hjorth, S.A., Bergsma, D., Sathe, G. and Schwartz, T.W. (1994). Differentiation between binding epitopes for angiotensin II and nonpeptide antagonists on the angiotesin II type 1 receptors. *Proc. Natl. Acad. Sci. USA,* **91**, 7046–7050.

Schutz, W. and Freissmuth, M. (1992). Reverse intrinsic activity of antagonists in G protein coupled receptors. *Trends Pharmacol. Sci.,* **13**, 376–380.

Schwartz, T.W. (1994). Locating ligand-binding sites in 7TM receptors by protein engineering. *Current Opinion Biotech.,* **5**, 434–444.

Shenker, A., Laue, L., Kosugi, S., Merendino, J.J., Minegishi, T. and Cutler, G.B. (1993). A constitutively activating mutation of the luteinizing hormone receptor in familial male precocious puberty. *Nature,* **365**, 652–654.

Stephenson, R.P. (1956). A modification of receptor theory. *J. Pharmacol. Chemother.,* **11**, 379–393.

CHARACTERIZATION OF STEREOSELECTIVE INTERACTIONS OF CATECHOLAMINES WITH THE α_{2a}-ADRENOCEPTOR *VIA* SITE DIRECTED MUTAGENESIS

J. PAUL HIEBLE, ANDREAS HEHR, YING-OU LI, DIANE P. NASELSKY and ROBERT R. RUFFOLO, JR.

Division of Pharmacological Sciences, SmithKline Beecham Pharmaceuticals, 709 Swedeland Road, King of Prussia, Pennsylvania 19406

It has been established that the amine nitrogen of the catecholamine neurotransmitters forms an ionic bond with the carboxyl group on the side chain of an aspartic acid residue located in transmembrane helix III of α- and β-adrenoceptors (Asp[113] in the β_2- and α_{2a}-adrenoceptors). While the specific sites of interaction may vary, the ring aromatic hydroxyl groups appear to interact *via* hydrogen bonding to serine or cysteine residues of transmembrane helix V. Less is known regarding the site of interaction of the β-hydroxyl group of the catecholamines with the adrenoceptors. Functional studies have shown that the orientation of this hydroxyl group, located on the asymmetric β-carbon atom, is a key determinant of receptor affinity. Studies with the β_2-adrenoceptor have suggested several potential binding sites for the β-hydroxyl group, including Ser[165] (Helix IV), Asn[293] (Helix VI), Ser[319] (Helix VII) and Tyr[329] (Helix VII). However, most of these putative interactions are based only on molecular modeling, without supporting/confirming data from site-directed mutagenesis.

In the human recombinant α_{2a}-adrenoceptor, mutation of Ser[90], located on helix II, to alanine results in a marked reduction (50–70 fold) in the affinity of R(–)-norepinephrine and R(–)-epinephrine. This mutation has little effect ($\leq$ 4 fold reduction in affinity) on the ability of the respective S(+)-enantiomers, or their corresponding desoxy analogs to inhibit the binding of [^{3}H] clonidine to COS cells transiently expressing the mutant receptor. A selective reduction in the affinity of the R(–)-enantiomers of the catecholamines was also observed when Ser[419] (corresponding to Ser[319] of the β_2-adrenoceptor) was replaced by alanine. However, a similar mutation at Ser[165] had virtually no effect on the affinity of the α_{2a}-adrenoceptor for either the R(–)- or S(+)-enantiomers of the catecholamines. The α_{2a}-adrenoceptor contains three adjacent amino acids capable of hydrogen bonding in the region corresponding to Asn[293] of the β_2-adrenoceptor. Mutation to remove all three of these potential sites of agonist-receptor interaction selectively reduces the affinity of the R(–)-catecholamines. However, the magnitude and selectivity of this affinity reduction is substantially less than observed for the Ser[90] and Ser[419] mutations.

These data show that binding of catecholamine agonists to the α_{2a}-adrenoceptor does not involve Ser[165] in transmembrane helix IV. The selective reduction in affinity of the R(–)-catecholamines by the other mutations suggests that the β-hydroxyl group may interact with the α_{2a} adrenoceptor at multiple sites.

INTRODUCTION

Since the initial introduction of the basic concepts of chemical neurotransmission seventy years ago, there has been speculation regarding the mode of interaction between neurotransmitters and their receptors. In 1933, Easson and Stedman postulated the hypothesis

---

Correspondence to: J. Paul Hieble, Ph.D., Pharmacological Sciences, UW2510, SmithKline Beecham Pharmaceuticals, 709 Swedeland Road, P.O. Box 1539, King of Prussia, PA 19406-0939, Phone: (610) 270-6053, Fax: (610) 270-5080

that the active enantiomers of the catecholamines (i.e., (–)-enantiomers) bound to their receptors *via* a three point attachment, occurring through the amine nitrogen, ring hydroxyl groups and aliphatic β-hydroxyl group. Only in the active R(–)- conformation would these three groups all be in the proper orientation for binding to the receptor. The inactive S(+)- configuration, as well as the achiral analogs lacking the β-hydroxyl group (i.e. desoxy analogs) bound through only two sites with a consequent reduction in affinity. Since this proposal, the Easson-Stedman Hypothesis, as applied to the interaction of phenethylamines with α- and β-adrenoceptors, has been supported by data in many model systems showing the potency order R(–) >> S(+) = Desoxy (Nichols and Ruffolo, 1991).

The primary amino acid sequences of the α- and β-adrenoceptors are now known, and the sites of interaction between agonist and receptor can be studied by a variety of techniques, including site-directed mutagenesis and molecular modeling. Site directed mutagenesis has been applied extensively to the $β_2$-adrenoceptor, and has been used to show convincingly that the amino nitrogen and ring hydroxyl groups interact with Asp^{113} (Transmembrane Helix III) and Ser^{204}, Ser^{207} (Transmembrane Helix V), respectively (Strader *et al.*, 1994).

The role of these amino acid residues in the binding of catecholamines to the β-adrenoceptor are supported by molecular models created of the $β_2$-adrenoceptor. Although the relative orientation of the transmembrane helices can vary substantially between specific models (Trumpp-Kallmeyer *et al.*, 1992; Maloney-Huss and Lybrand, 1992; Venter *et al.*, 1989), they all can accommodate the binding of norepinephrine or epinephrine between Helices III and V.

Despite the importance of the aliphatic β-hydroxyl group in the stereoselective binding of catecholamines to their receptors, only a few reports have addressed the potential binding sites for this critical functional group. Based on molecular modeling of the $β_2$-adrenoceptor, Ser^{165} in transmembrane helix IV has been suggested to be a likely site for this interaction (Strosberg, 1993; Trumpp-Kallmeyer *et al.*, 1992). However, when this serine residue was replaced by alanine, no functional $β_2$-adrenoceptor was expressed (Strader *et al.*, 1989), preventing the direct testing of this hypothesis. Another molecular model suggests binding of the aliphatic hydroxyl group to Ser^{319} in transmembrane helix VII (Maloney-Huss and Lybrand, 1992). Replacement of this serine residue by alanine results in a 15-fold loss in the potency of isoproterenol as a stimulant of adenylate cyclase, and a reduction in the maximal stimulation observed (Dixon *et al.*, 1988). This mutation also resulted in a tenfold-reduction in the ability of isoproterenol to inhibit the binding of $[^{125}I]$ ICYP to the $β_2$-adrenoceptor without influencing the affinity of alprenolol, a β-adrenoceptor antagonist (Strader *et al.*, 1989). The stereoselectivity of the reduction in isoproterenol affinity in these assays was not determined.

Based on molecular modeling of the $β_2$-adrenoceptor which suggested that Asn^{293} in transmembrane helix VI might interact with the β-hydroxyl group of isoproterenol, Wieland *et al.* (1994) replaced this residue by leucine and aspartic acid. The degree of stereoselectivity observed for the enantiomers of isoproterenol was reduced by mutation of Asn^{293} to leucine. Replacement of Asn^{293} by aspartic acid had no effect on the stereoselectivity of isoproterenol.

Another potential site for interaction between the catecholamine β-hydroxyl group and the $β_2$-adrenoceptor, Tyr^{329} in transmembrane helix VII, has been proposed based on molecular modeling studies (Strosberg, 1995). No mutagenesis data to support an interaction with this residue has been reported to date.

Much less information is available regarding the interactions of catecholamines and the α_2-adrenoceptors. As with the β_2-adrenoceptor, the amino nitrogen of the catecholamines appears to interact with Asp^{113} (Transmembrane Helix III) of the α_{2a}-adrenoceptor (Wang et al., 1991). Likewise, there is evidence to support an interaction of the *para*-hydroxyl group on the aromatic ring with Ser^{204} (corresponding to Ser^{207} of the β_2-adrenoceptor) in transmembrane helix VI. Site-directed mutagenesis did not identify a specific residue binding the *meta*-hydroxyl group, although binding to Cys^{201} was postulated (Wang et al., 1991; Lee et al., 1994). The only report suggesting a binding site for the catecholamine β-hydroxyl group to the α_2-adrenoceptor was the molecular modeling study of Trumpp-Kallmeyer et al. (1992) which postulated that, as in the β_2-adrenoceptor, this group interacts with Ser^{165} in transmembrane helix IV.

We have previously reported that replacement of Ser^{90} in transmembrane helix II of the α_{2a}-adrenoceptor by alanine results in a selective reduction in the affinity of the R(–)-enantiomers of the catecholamines (Li et al., 1995), suggesting that this site may be involved in the interaction with the β-hydroxyl group. We now compare the effects of mutation at this site with corresponding mutations of several other serine residues, all of which have been postulated to be involved in the interaction of the catecholamine β-hydroxyl group with the β_2-adrenoceptor. These include Ser^{165} (Transmembrane Helix IV) and Ser^{419} (Transmembrane Helix VII, corresponding to Ser^{319} of the β_2-adrenoceptor). In addition, we evaluated an α_{2a}-adrenoceptor mutant where Thr^{393}, Tyr^{394} and Thr^{395}, located at the site on transmembrane helix VI corresponding to Asn^{293} of the β_2-adrenoceptor, were replaced by alanine, phenylalanine and alanine, respectively.

METHODS

α_{2a}-adrenoceptor

A 4.5 kb cDNA clone encoding the human α_{2a}-adrenoceptor was isolated from a human hippocampus cDNA library using as hybridization probes a series of degenerate synthetic oligonucleotides homologous to the seven transmembrane spanning domains of serotonin receptors. A 1.4 kb fragment encompassing the entire coding region of the cDNA (450 amino acids) was enzymatically excised from the original 4.5 kb clone and subcloned into Eco RI/Bam HI restriction sites of the pALTER plasmid vector for site-directed mutagenesis.

Preparation of mutant receptors

Site-directed mutagenesis was performed according to manufacturers directions (Altered Sites Mutagenesis System, Promega). The mutant gene was then inserted into the expression vector, pCDN (Aiyar et al., 1994), using the EcoRI and BamHI sites. Four mutant receptors were prepared: $Ser^{90} \rightarrow$ Alanine; $Ser^{165} \rightarrow$ Alanine; $Ser^{419} \rightarrow$ Alanine and $Thr^{393} \rightarrow$ Alanine, $Tyr^{394} \rightarrow$ Phenylalanine, $Thr^{395} \rightarrow$ Alanine.

Transient expression in COS cells

COS cells were maintained in DMEM containing 10% fetal bovine serum (GIBCO) at 37°C in 5% CO_2/95% O_2. Cells were plated the day before transfection in multiple T150 flasks and transfected using the DEAE-dextran method (McCutchan and Pagano, 1968) with 40 μg DNA per flask. After 72 hours, cells were washed with BPS, pelleted, then frozen in a dry ice/ethanol bath, and stored at –70°C.

Establishment of stable CHO clone expressing wild type α_{2a}-adrenoceptor

Chinese hamster ovary cells, grown in suspension using MR1MOD3 medium (in-house formulation), were transfected by electroporation with DNA containing the gene encoding the α_{2a}-adrenoceptor. Cells were expanded and tested for the presence of receptor expression by binding to [^{3}H] rauwolscine.

Membrane preparation

Previously frozen cell pellets were thawed and homogenized in a Brinkman Polytron with cold TME buffer (50 mM Tris, 12.5 mM $MgCl_2$, 5 mM EDTA, pH 7.4). The homogenate was centrifuged at 100,000 × g for 30 minutes at 4°C. The membrane pellet was resuspended in TME buffer, pH 7.4.

Radioligand binding studies

Membranes were incubated with test compounds at concentrations ranging from 0.1 nM to 100 μM. Assays were initiated by the addition of membrane protein, and incubated at 25°C for 45 minutes. 50 mM Tris-EDTA buffer, pH 7.4, was used for [^{3}H] rauwolscine and [^{3}H] clonidine. For the [^{3}H] norepinephrine assay, 1 mM EDTA, 0.01 mM dithiothreitol, 0.1 M catechol and 0.1% ascorbic acid; furthermore, the Whatman GF/B filters used for sample separation were presoaked in a solution of 1 mM catechol dissolved in 0.1% ascorbic acid. Filters were rinsed 3 times with 5 ml of ice-cold wash buffer containing 50 mM Tris-HCl (pH 7.7) at 25°C, 0.1% ascorbic acid, and 1 mM catechol. Radioligands were present at concentrations near their K_d values (Tables 1 and 2). Non-specific binding was defined using 10 μM phentolamine. K_i values were calculated using the London software program for nonlinear regression.

TABLE 1
Binding of [^{3}H] clonidine to wild type and mutant α_{2a}-adrenoceptors, transiently expressed in COS cells

Receptor	K_d (nM)	B_{max} (pmol/mg)
Wild Type	2.6 ± 0.6	2.2 ± 0.1
Ser^{90} → Alanine	3.2 ± 0.8	5.9 ± 1.9
Ser^{419} → Alanine	12.7 ± 2.0	3.1 ± 0.6
Thr^{393} → Alanine Tyr^{394} → Phenylalanine Thr^{395} → Alanine	1.2 ± 0.1	5.1 ± 0.2
Ser^{165} → Alanine	3.6 ± 0.6	3.1 ± 0.3

TABLE 2
Effect of radioligand on the apparent affinity of rauwolscine, clonidine and norepinephrine for α_{2a}-adrenoceptors expressed in CHO cells

Radioligand	K_d (nM)	K_i (Rauwolscine)	K_i (Clonidine)	K_i (Norepinephrine)
[^{3}H] Rauwolscine	1.0	3.5 ± 0.5	26.2 ± 2.7	996 ± 63
[^{3}H] Clonidine	1.0	4.4 ± 1.1	1.5 ± 0.3	7.1 ± 1.2
[^{3}H] Norepinephrine	13.0	3.8 ± 0.8	9.6 ± 1.6	15.4 ± 3.3

RESULTS

As has been previously reported, R(−)-norepinephrine was a weak inhibitor of the binding of [^{3}H] rauwolscine to the recombinant α_{2a}-adrenoceptor. Although mutation of Ser^{90} to alanine produced a further reduction in the ability of R(−)-norepinephrine to inhibit the binding of this radioligand (data not shown), the low affinity of the catecholamines for the wild-type α_{2a}-adrenoceptor in this assay complicated the quantitation of the effect of site-directed mutagenesis on receptor affinity. For this reason, we evaluated α_2-adrenoceptor affinity using [^{3}H] clonidine as the radioligand. The K_i value for R(−)-norepinephrine as an inhibitor of the binding of [^{3}H] clonidine is much lower than that observed against [^{3}H] rauwolscine, and comparable to the affinity of norepinephrine for the α_{2a}-adrenoceptor as measured by the K_d for its interaction with [^{3}H] norepinephrine, or the K_i for the inhibition of [^{3}H] norepinephrine by unlabeled norepinephrine (Table 2).

Table 3 shows that mutation of Ser^{90} (transmembrane helix II) or Ser^{419} (transmembrane helix VII) of the α_{2a}-adrenoceptor to alanine produces a 36–73 fold reduction in the affinity of the receptor for the active enantiomers of the catecholamine neurotransmitters. The inactive S(+)-enantiomers had 50–100 fold lower affinity for the wild type receptor (Table 3); however, mutation of Ser^{90} or Ser^{419} produced only a 2–4 fold reduction in the affinity of the S(+)-catecholamines for the α_{2a}-adrenoceptor. Likewise, these mutations had little effect on the affinity of dopamine or epinine, the des-hydroxy analogs of the catecholamine neurotransmitters.

In contrast to the effects of mutation of Ser^{90} or Ser^{419}, mutation of Ser^{165} had virtually no effect on the affinity of the α_{2a}-adrenoceptor for any of the catecholamines (Table 3).

Mutation of three adjacent amino acids in transmembrane helix VI (Thr^{393}, Tyr^{394}, Thr^{395}) to eliminate hydroxyl groups capable of interaction *via* hydrogen bonding to the β-hydroxyl group of norepinephrine and epinephrine reduced the affinity of the α_{2a}-adrenoceptor for both R(−)- and S(+)-enantiomers (Table 3). Although the affinity reduction was greater for the R(−)-enantiomers, the magnitude of the reduction was less than that observed for mutation of either Ser^{90} or Ser^{419}; furthermore the mutation in transmembrane helix VI produced a greater reduction in the affinity of the S(+)-enantiomers, so that the stereospecificity of the affinity reduction was substantially less.

DISCUSSION

The Easson-Stedman Hypothesis predicts that the catecholamines interact with their receptors with the potency order of R(−) > S(+) = Desoxy. While this potency order is

TABLE 3
Determination of the effect of site-directed mutagenesis on the affinity of the α_{2a}-adrenoceptor for catecholamines

Compound	K_i (WT)[a]	K_i (Ser^{90})[b]	Ratio	K_i (Ser^{165})[b]	Ratio	K_i (Ser^{419})[b]	Ratio	K_i (Thr^{393}, Tyr^{394}, Thr^{395})[b]	Ratio
(R) Norepinephrine	31 ± 6	1522 ± 619	49	58 ± 10	1.9	1221 ± 194	39	361 ± 78	12
(S) Norepinephrine	3549 ± 800	4512 ± 493	1.3	2166 ± 490	0.6	6756 ± 636	1.9	16,242 ± 3638	4.6
Dopamine	329 ± 69	668 ± 99	2.0	138 ± 67	0.4	553 ± 71	1.7	602 ± 167	1.8
(R) Epinephrine	13 ± 5	958 ± 141	73	36 ± 14	2.7	477 ± 99	36	258 ± 83	20
(S) Epinephrine	687 ± 93	2988 ± 1123	4.3	1486 ± 693	2.2	3124 ± 236	4.5	5845 ± 2139	8.5
Epinine	360 ± 139	1127 ± 380	3.1	ND	ND	982 ± 93	2.7	ND	ND

[a]K_i (nM) for inhibition of [^{3}H] clonidine binding to COS cells transiently expressing wild type or mutant α_{2a}-adrenoceptors
[b]Site of mutation. For amino acid in mutated receptor see Table 1.

commonly observed in functional tissue assays (Nichols and Ruffolo, 1991), radioligand binding assays often show, as in our experiments, that dopamine has an affinity intermediate between the more active and less active enantiomer of norepinephrine. This would suggest that the β-hydroxyl group of norepinephrine results in a favorable interaction with the receptor when in the R(–) configuration, and an unfavorable interaction when in the S(+) configuration. It is possible that for (S) norepinephrine, the β-hydroxyl group, in an improper orientation for interaction with its binding site, now interacts sterically with another portion of the receptor molecule to prevent optimal interaction of the amino group and ring hydroxyl groups with their respective binding sites.

The favorable interaction of the β-hydroxyl group with the receptor is reflected in the tenfold greater affinity of the wild type α_{2a}-adrenoceptor for R(–) norepinephrine compared to dopamine (Table 3). In contrast, both enantiomers of norepinephrine have lower affinity than dopamine for the mutant α_{2a}-adrenoceptors where Ser^{90} or Ser^{419} is replaced by alanine. Hence these mutations convert a favorable interaction between β-hydroxyl group and receptor to an unfavorable interaction.

There are several possible interpretations of these data: 1) The β-hydroxyl group of catecholamines may be capable of interaction with either transmembrane helices II and VII, so that elimination of either site of hydrogen bonding reduces affinity, 2) Ser^{90} and Ser^{419} interact with each other, through intervening water molecules present in the receptor cavity such that alteration of either residue would disrupt this interaction and result in a change in receptor conformation, or 3) both Ser^{90} and Ser^{419} are independently involved in interactions with other portions of the receptor in order to maintain the cavity in a conformation optimal for binding of catecholamines. It is also possible that Ser^{419}, which is conserved throughout most of the receptors for biogenic amines, plays a structural role in maintaining receptor conformation, while Ser^{90}, conserved only in the α-adrenoceptors and most 5-HT receptors, is involved in a specific interaction with the catecholamine β-hydroxyl group.

It is possible that conformational changes in the receptor cavity could result in a selective reduction in affinity for the R(–) catecholamines, since these are the only agents predicted to interact with the receptor via a three point attachment. This three point attachment, predicted by the Easson-Stedman Hypothesis to be responsible for the high affinity of these molecules, could be more sensitive to subtle changes in receptor geometry than the two point attachment predicted for other α-adrenoceptor agonists. However, our observation that mutation of Ser^{90} or Ser^{419} does not substantially reduce the affinity of the α_{2a}-adrenoceptor for the S(+)-or desoxy catecholamines, or for a variety of agonists, including imidazolines, aminotetralins, and phenethylamines lacking ring-hydroxyl groups (Li et al., 1995; Hieble, unpublished data) would suggest that the geometry of the receptor cavity is not influenced by these mutations.

As shown in Table 3, replacement of Ser^{165} in transmembrane helix IV by alanine has virtually no effect on the affinity of the α_{2a}-adrenoceptor for the catecholamines. In contrast to the results obtained for the β_2-adrenoceptor, where mutation of Ser^{165} was incompatible with the expression of a functional receptor (Strader et al., 1989), the α_{2a}-adrenoceptor with alanine at this position still shows stereoselective interaction with the catecholamines. Although molecular modeling has suggested that Ser^{165} is the binding site for the β-hydroxyl group in both α- and β-adrenoceptors (Trumpp-Kallmeyer et al., 1992), our data shows that this is not the case for the α_{2a}-adrenoceptor.

Mutation of Asn^{293} in transmembrane helix VI of the β_2-adrenoceptor to leucine has been found to reduce the stereoselectivity of the interaction of isoproterenol with the receptor, suggesting that this asparagine residue is participating in the binding of the β-hydroxyl group (Wieland *et al.*, 1994). However, it was not reported whether this loss in stereoselectivity results from a decreased affinity for the R(–)-enantiomer or, as we have found for some other mutations of the α_{2a}-adrenoceptor (Hehr *et al.*, 1996), an increase in affinity for the "inactive" S(+)-enantiomer. The α_{2a}-adrenoceptor contains three adjacent sites of potential hydrogen bonding in transmembrane helix VI at a position corresponding to Asn^{293} in the β_2-adrenoceptors. Site directed mutagenesis to remove all three of these potential sites of interaction with the catecholamines results in a 12–20 fold reduction in the affinity of the R(–)-enantiomers (Table 3). Although the magnitude of this affinity reduction was greater than observed for the S(+)-enantiomers (4.6–8.5 fold), the reduction in affinity for the S(+)-enantiomers was greater than observed for the other mutations. It is possible that a non-specific reduction in affinity is produced as a consequence of mutation of these three amino acids, two of which may not be involved in a specific interaction between catecholamine and receptor. Further experiments are underway to determine the specific change responsible for the reduction in affinity for the R(–) catecholamines.

Although we have not yet identified a single unique site for interaction of the aliphatic β-hydroxyl group of the catecholamines with the α_{2a}-adrenoceptor, site-directed mutagenesis has shown that in contrast to predictions based on molecular modeling, Ser^{165} on transmembrane helix IV does not appear to be involved in the binding of the neurotransmitter molecules. Furthermore, Ser^{90} (transmembrane helix II) and Ser^{419} (transmembrane helix VII) are likely to be involved in the stereoselective interaction of catecholamines with the α_{2a}-adrenoceptor, most probably by providing a binding site for the β-hydroxyl group.

References

Aiyar, N., Bayer, E., Wu, H.-L., Nambi, P., Edwards, R.M., Trill, J.J., Ellis, C. and Bergsma, D.J. (1994). Human AT-1 receptor is a single copy gene: Characterization in a stable cell line. *Mol. and Cell. Biochem.*, **131**, 75–86.

Dixon, R.A.F., Sigal, I.S. and Strader, C.D. (1988). Structure-function analysis of the β-adrenergic receptor. *Cold Spring Harbor on Quantitative Biology*, **53**, 487–497.

Easson, L.H. and Stedman, E. (1933). Studies on the relationship between chemical constitution and physiological action. V. Molecular dyssemetry and physiological activity. *Biochem. J.*, **27**, 1257–1266.

Hehr, A., Hieble, J.P., Li, Y.-O., Bergsma, D.J., Swift, A.M., Ganguly, S. and Ruffolo, R.R., Jr. (1996). Ser^{165} of transmembrane helix IV is not involved in the interaction of catecholamines with the α_{2a}-adrenoceptor. *Pharmacol.*, in press.

Lee, N.H., Pellegrino, S.M. and Fraser, C.M. (1994). Site-directed mutagenesis of α-adrenergic receptors. *Neuroprotocols*, **4**, 20–31.

Li, Y.-O., Hieble, J.P., Bergsma, D., Swift, A.M., Ganguly, S. and Ruffolo, R.R., Jr. (1995). The β-hydroxyl group of catecholamines may interact with Ser^{90} of the second transmembrane helix of the α_{2a}-adrenoceptor. *Pharmacol. Comm.*, **6**, 125–131.

Maloney-Huss, K. and Lybrand, T.P. (1992). Three-dimensional structure for the β_2-adrenergic receptor protein based on computer modeling studies. *J. Mol. Biol.*, **225**, 859–871.

McCutchan, J.H. and Pagano, J.S. (1968). Enhancement of the infectivity of simian virus 40 deoxyribonucleic acid with dethylaminoethyl-dextran. *J. Natl. Cancer Inst.*, **41**, 351–357.

Nichols, A.J. and Ruffolo, R.R., Jr. (1991). Structure activity relationships for α-adrenoceptor agonists and antagonists. In: α-*Adrenoceptors: Molecular Biology, Biochemistry and Pharmacology. Prog. Basic. Clin. Pharmacol.*, **8**, 75–114. Basel, Karger.

Strader, C.D., Candelore, M.R., Hill, W.S., Sigal, I.S. and Dixon, R.A.F. (1989). Identification of two serine residues involved in agonist activation of the β-adrenergic receptor. *J. Biol. Chem.*, **264**, 13572–13578.

Strader, C.D., Fong, T.M., Tota, M.R. and Underwood, D. (1994). Structure and function of G protein coupled receptors. *Annu. Rev. Biochem.*, **63**, 101–132.

Strosberg, A.D. (1993). Structure, function and regulation of adrenergic receptors. *Protein Science*, **2**, 1198–1209.

Strosberg, A.D. (1995). Structure, function and regulation of the three β-adrenoceptor subtypes. In: *G-Protein Coupled Transmembrane Signaling Mechanisms*, R.R. Ruffolo, Jr. and M.A. Hollinger, eds. CRC Press, Boca Raton, pp. 35–56.

Trumpp-Kallmeyer, S., Hoflack, J., Bruinvels, A. and Hibert, M. (1992). Modeling of G-protein-coupled receptors: Application to dopamine, adrenaline, serotonin, acetylcholine and mammalian opsin receptors. *J. Med. Chem.*, **35**, 3448–3462.

Venter, J.C., Fraser, C.M., Kerlavage, A.R. and Buck, M.A. (1989). Molecular biology of adrenergic and muscarinic cholinergic receptors. *Biochem. Pharmacol.*, **38**, 1197–1208.

Wang, C.D., Buck, M.A. and Fraser, C.M. (1991). Site-directed mutagenesis of α_{2a}-adrenergic receptors: Identification of amino acids involved in ligand binding and receptor activation by agonists. *Mol. Pharmacol.*, **40**, 168–179.

Wieland, K., Ijzerman, A.P. and Lohse, M.J. (1994). Involvement of asparagine-292 in the human β_2-adrenergic receptor in stereospecific agonist binding. *Naunyn-Schmiedeberg's Arch. Pharmacol.*, **349 (Suppl.)**, R10.

REGULATING SIGNAL TRANSFER FROM RECEPTOR TO G-PROTEIN

MOTOHIKO SATO, GUANGYU WU and STEPHEN M. LANIER

Department of Pharmacology, Medical University of South Carolina, Charleston, SC USA

Members of the G-protein coupled receptor superfamily play a central role in cellular communication mediating the cell response to numerous hormones and neurotransmitters. Via coupling to heterotrimeric guanine nucleotide binding proteins (G), receptor activation regulates numerous intracellular effector molecules including ion channels, adenylyl cyclases and phospholipases. Many of these receptors share common signalling pathways that are activated with varying degrees of efficiency by a particular hormone. The specific intracellular signal initiated by agonist and the strength of signal propagation likely depend on many factors including the relative types and/or amounts of the signal transducing entities [receptor (R)/G-protein/effector (E)]. Additional determinants of signalling specificity include cell architecture and perhaps accessory proteins that regulate events at the R-G or G-E interface.

In attempts to further define determinants of signalling specificity, we focused our efforts on one receptor subfamily, α_2-adrenergic receptors (α_2-AR). Utilizing the α_2-AR subfamily as a representative subgroup of G-protein coupled receptors, we have attempted to define how signalling specificity is engineered in the intact cell. Several approaches to this isssue were employed. One effort focused on the development of strategies to identify proteins other than G-protein that might interact with the receptor protein. A second approach entailed the partial purification and characterization of proteins that may regulate the transfer of signal from R to G.

KEY WORDS: G-protein, cell signal, adrenergic receptor, epinephrine

INTRODUCTION

Cells sense and respond to their environments via one or more innovative and clever signal processing systems. The majority of such systems involve molecules on the cell surface that respond to external stimuli with some sort of conformational change initiating a complex series of events inside the cell leading to a specific response to the external stimuli (Figure 1). Such a response might be contraction, relaxation, growth arrest or secretion depending upon the cell type and particular stage of cell development. As one might imagine, dysfunction or miscues in these signalling events can severely compromise the normal functioning of the cell.

Our efforts are focused on dissecting some of these pathways to define the precise molecular interactions involved in signal propagation. Specifically we are interested in the superfamily of membrane receptors (R) that couple to heterotrimeric G-proteins (G) (Bourne *et al.*, 1991; Hepler and Gilman, 1992; Neer, 1995; Sato *et al.*, 1995). One of

---

Correspondence to: Dr. Stephen Lanier, Department of Cell and Molecular Pharmacology and Experimental Therapeutics, Medical University of South Carolina, 171 Ashley Avenue, Charleston, SC 29425. Tel (803)792-2574, Fax (803)792-2475, email laniersm@musc.edu

EXTERNAL STIMULI

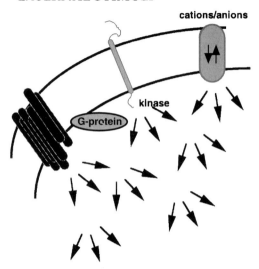

SPECIFIC RESPONSE TO STIMULI

FIGURE 1 Signal processing by the cell. Refer to text for details.

the central questions in this field is as follows: *What are the interactions between isoforms or subtypes of the signalling molecules that account for signalling specificity in various cell types at specific stages of development?* Utilizing the α_2-adrenergic receptor (α_2-AR) subfamily of G-protein coupled receptors, we have attempted to address this question. Our experimental approach initiated several years ago involved the expression of the receptor gene in different environments — i.e. different cell phenotypes, followed by analysis of receptor coupling events (Sato *et al.*, 1995; Lanier *et al.*, 1991; Duzic *et al.*, 1992; Duzic and Lanier, 1992; Coupry *et al.*, 1992; Hamamdzic *et al.*, 1995). Several observations lead us to suggest that there are accessory proteins that regulate events at the R-G or G-E interface and participate in the generation of a signal transduction complex on the inner face of the membrane. As an initial approach to identify components of this putative signal transduction complex we used two experimental systems. One involves the use of receptor subdomain probes to search for interacting proteins. The second approach involves a solution phase assay to search for proteins that directly activate G.

EXPERIMENTAL PROCEDURES

Materials

[^{3}H]RX821002 (60 Ci/mmol) and [^{32}P]γATP (3000 Ci/mmol) were obtained from Amersham. [^{35}S]GTPγS and [^{3}H]UK14304 were purchased from Dupont/NEN (Boston, MA). Glutathione Sepharose 4B was purchased from Pharmacia Biotech. Tissue culture supplies were obtained from JRH Bioscience (Lenexa, KS). Acrylamide, bis-acrylamide and SDS were purchased from Bio-Rad (Hercules, CA). Nitrocellulose and PVDF membranes were obtained from Schleicher & Schuell (Keene, NH). Propranolol, epinephrine and pertussis toxin were obtained from Research Biochemicals Inc (Natick, MA). Ecoscint A was purchased from National Diagnostics (Manville, NJ). Guanosine diphosphate, thesit (polyoxyethylene-9-lauryl ether) and guanyl-thiodiphosphate were obtained from Boehringer-Mannheim (Indianapolis, IN).

Membrane/G-protein reconstitution

Membranes were prepared from NIH-3T3 fibroblasts and PC-12 pheochromocytoma cells stably transfected with the rat RG-20 $\alpha_{2A/D}$-AR as previously described (Duzic and Lanier, 1992). In this assay system, receptor coupling to endogenous G-proteins was eliminated by pretreatment of cells with pertussis toxin and the receptor-mediated signal was reconstituted in pertussis toxin treated membranes by addition of G-protein purified from bovine brain. Thus the receptor has access to same population of G-proteins (Sato *et al.*, 1995). Briefly, the assay system was as follows. The preincubation cocktail consisted of brain G-protein and membrane in buffer A (50 mM Tris-HCl pH 7.4, 5 mM $MgCl_2$, 0.6 mM EDTA) containing 0.005% thesit and GDP. The concentrations of brain G-protein and GDP in the preincubation cocktail were five times the final concentration desired in the assay tube. After one hr incubation of the preincubation cocktail at 4°C, 20 μl of the cocktail was added to the assay tube containing buffer A plus (final concentrations) 150 mM NaCl, 1 mM DTT, 1 μM propranolol, ~50,000 cpm (~0.2 nM) [^{35}S]GTPγS and agonist, vehicle or 100 μM GTPγS (total vol = 80 μl). Incubation was then continued at 24°C for various times and the reaction terminated by rapid filtration through nitrocellulose filters (Schleicher & Schuell BA85) with buffer B (100 mM NaCl, 50 mM Tris-HCl 5 mM $MgCl_2$ pH 7.4, 4°C). Radioactivity bound to the filters was determined by liquid scintillation counting. High-affinity, Gpp(NH)p-sensitive binding of the α_2-AR agonist [^{3}H]UK14304 was determined in a similar manner with exclusion of sodium and GDP from the preincubation cocktail. Nonspecific binding was defined with 100 μM GTPγS or Gpp(NH)p.

Analysis of cytosolic proteins interacting with receptor subdomains

The receptor subdomain encoding residues K318-G364 of the $\alpha_{2A/D}$-AR third intracellular loop was amplified by polymerase chain reaction (PCR) and the product was cloned into the BamH1 and EcoR1 restriction sites of the expression vector pGEX-2T and pGSTag. GST fusion proteins were then expressed in the E. coli strain HB101. Fusion protein synthesis was induced by isopropyl-β-D-thiogalactopyranoside (IPTG) and the fusion protein was purified using a glutathione-sepharose affinity resin as described by Smith and Johnson (1988). [^{32}P]-labeled receptor subdomain probes were generated by thrombin cleavage of resin-bound fusion protein following phosphorylation with [^{32}P]γATP and protein kinase A (Ron and Dressler, 1992). Cytosolic proteins from bovine brain were first fractionated by ammonium sulfate precipitation, and each fraction further separated by DEAE ion exchange chromatography. Fractions of enriched brain cytosol proteins were screened for interacting proteins with a receptor subdomain by gel overlay as described by Ron and Dressler (1992) and by affinity chromatography as described elsewhere (Xu *et al.*, 1995).

RESULTS

To address the role of the receptor's microenvironment in determining signalling specificity/efficiency, we expressed α_2-adrenergic receptor subtypes in different cell phenotypes and evaluated signal propagation initiated by agonist (Sato et al., 1995; Lanier et al., 1991; Duzic et al., 1992; Duzic and Lanier, 1992; Hamamdzic et al., 1995). Initially, we focused on the ability of the receptor to increase the levels of the second messengers cAMP or calcium within the cell. In PC-12 receptor transfectants, receptor activation augments the effects of cyclase catalytic stimulators on cellular cAMP and elicits an increase in intracellular calcium (Duzic and Lanier, 1992). These actions are not observed in NIH-3T3 receptor transfectants. The augmentation of the forskolin-induced increase in cAMP following receptor activation in PC-12 cells requires the intact cell and is dependent upon a simultaneous increase in intracellular calcium. The observed effect is actually further exaggerated after eliminating the coupling of the receptor to G_i or G_o by cell pretreatment with pertussis toxin. The positive coupling to adenylylcyclase (AC) likely involves a pertussis toxin insensitive G-protein that increases intracellular calcium which perhaps facilitates the effect of forskolin on a calcium/calmodulin regulated isoform of AC. Indeed, analysis of mRNA from the three cell types by RT-PCR and RNA blot analysis indicated the expression of the Type III AC transcript in PC-12 cells but not NIH-3T3 fibroblasts or DDT_1-MF_2 cells (Sato, M., Duzic, E. and Lanier, S.M., unpublished data). Type III AC is similar to type I and VIII enzymes, which are regulated by calcium/calmodulin.

Subsequently, we moved more proximal in the signal transduction cascade and examined the transfer of signal from R to G in NIH-3T3, PC-12 and DDT_1-MF2 $\alpha_{2A/D}$-AR

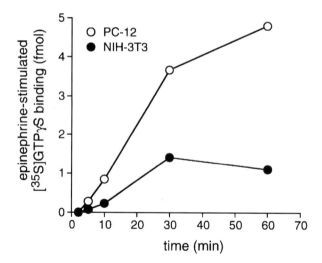

FIGURE 2 Agonist-induced activation of G-protein in NIH-3T3 and PC12 transfectants. Membranes were prepared from cells pretreated with pertussis toxin and reconstituted with 25 nM bovine brain G-protein as described in "Experimental Procedures". The binding of [^{35}S]GTPγS to the reconstituted preparation was determined in the presence or absence of epinephrine (10 μM) for various reaction times at 24°C. The results are presented as the mean of duplicate determinations and are representative of three experiments using different membrane preparations. GDP = 10 μM. Receptor density (fmol/mg membrane protein): NIH-3T3, 3800; PC-12, 3900. basal [^{35}S]GTPγS binding at 2, 5, 10, 30 and 60 min; NIH-3T3 0.07, 0.17, 0.34, 0.54, 0.87; PC-12 0.42, 0.46, 0.64, 1.24, 2.51.

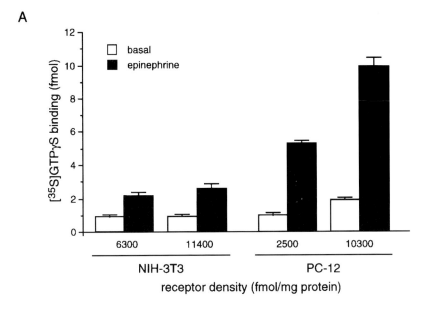

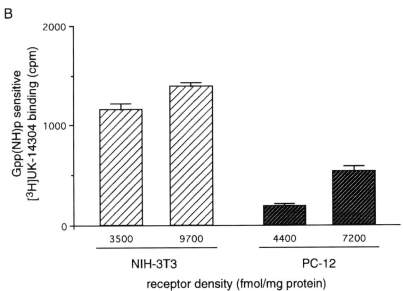

FIGURE 3 Comparison of agonist-induced activation of G-protein and Gpp(NH)p-sensitive binding of the selective α_2-AR agonist [^{3}H]-labeled UK14304 in NIH-3T3 and PC-12 $\alpha_{2A/D}$-AR transfectants in the membrane/G-protein reconstitution assay. Membranes were prepared from pertussis toxin treated cells and reconstituted with 25 (A) or 300 (B) nM brain G-protein as described in "Experimental Procedures". Results are presented as the mean ± SEM of four experiments. A) agonist-mediated effects on [^{35}S]GTPγS binding. Epinephrine = 10 μM. B) high-affinity agonist binding. [^{3}H]-labeled UK14304 = 1 nM. Gpp(NH)p = 100 μM. Data are presented as the total amount of specific ligand binding that is inhibited by Gpp(NH)p. Specific binding of [^{3}H]UK-14304 (1 nM) in the absence of Gpp(NH)p, cell, receptor density (fmol/mg protein): 1300.0 ± 81.5 cpm, NIH-3T3, 3500; 1594.0 ± 30.2 cpm, NIH-3T3, 9700; 321 ± 11.2 cpm, PC-12, 4400; 745.0 ± 79.5 cpm, PC12, 7200. Counting efficiency = ~50%. (Modified from Sato et al. (1995). J. Biol. Chem., **270**, 15269–15276.)

transfectants. We measured two events, 1) the population of receptors exhibiting high affinity for agonist and 2) the activation of G as reflected by guanine nucleotide exchange. The latter event is much more dynamic when the receptor is operating in the microenvironment found in the PC-12 neuronal like cell (Sato *et al.*, 1995). This augmented transfer of signal also occurs in a signal reconstitution system when the receptor is accessing the same population of G-proteins (Figures 2, 3). The percentage of the receptor population exhibiting high affinity for agonist is actually less in PC-12 cells suggesting that there is a smaller population of such receptors in PC-12 cells that transfer the signal to G with a greater efficiency. Alternatively, the greater degree of G-activation observed in PC-12 cells may result in a lower population of receptors exhibiting high affinity for agonist. There are several possible explanations for the observed differences in signal transfer. Perhaps the receptor is postranslationally modified in a cell-type specific manner that influences signal propagation. One might also imagine various types of cell-specific accessory proteins that stimulate or inhibit the transfer of signal from R to G (black figurines in Figure 5). Such signal regulators may also act directly on G. Initial experiments suggest that there is a membrane associated protein in the neuronal like cells PC-12 and NG108-15 that directly activates G. This and other efforts has led to the working hypothesis that the signalling specificity/efficiency is determined in part by proteins other than R, G and E found in the receptor's microenvironment (Figure 5). Such accessory proteins may participate in the formation of a signal transduction complex on the inner face of the membrane and serve to localize the signalling molecules and influence signal propagation.

In attempts to identify components of this putative signal transduction complex, we are utilizing both the functional assay described above and a strategy based upon protein-protein interactions. In the latter approach, we generated receptor subdomain probes from the third intracellular loop of the $\alpha_{2A/D}$-AR. Such probes were synthesized in bacteria as GST fusion proteins using the pGEX-2T or pGSTag vectors and used for gel overlay or affinity chromatographic analysis of brain cytosol (Figure 4). The third intracellular loop of the α_2-AR is ~150 amino acids in length and the juxtamembrane regions of this loop are important for G-protein coupling. There are also sites for phosphorylation in this region by G-protein coupled receptor kinase, but the function of the remainder of this loop is unclear. Initially we used a 47 amino acid receptor subdomain (K318-G364) derived from the carboxy terminal portion of the third intracellular loop in the protein interaction assays. In addition to the thrombin cleavage site inserted between GST and the receptor subdomain, pGSTag also contained a consensus site for phosphorylation by protein kinase A (PKA). The fusion proteins were generated in bacteria and purified on a glutathione affinity matrix. For generation of [^{32}P]-labeled receptor subdomains, the purified protein bound to the matrix was first labeled by PKA and then released from the resin by thrombin cleavage for use in gel overlay assays. In the second approach, the GST-receptor subdomain bound to the glutathione resin was used as an affinity matrix for purification of interacting proteins. Both approaches utilized a preparation of bovine brain cytosol in which various proteins were enriched by ammonium sulfate precipitation and ion exchange chromatography.

In the first approach the fractionated brain cytosol was electrophoresed on denaturing polyacrylamide gels and transferred to nitrocellulose. The proteins on the blot were gradually renatured and the blot was incubated with the [^{32}P]-labeled receptor subdomain

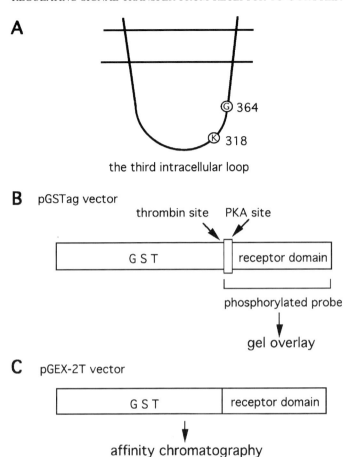

FIGURE 4 Construction of probes used for analysis of cytosolic proteins interacting with receptor subdomain. A) The third intracellular loop of the $\alpha_{2A/D}$-AR. Subdomain K318-G364 was used as a probe to screen for interacting proteins. B) Receptor subdomain K318-G364 was generated as GST fusion protein in pGSTag vector, liberated from GST by thrombin cleavage following phosphorylation by protein kinase A and used to identify interacting proteins by gel overlay. C) Subdomain K318-G364 was generated as GST fusion protein in pGEX-2T vector. Fusion proteins bound on glutathione-sepharose were used as affinity matrix to screen for interacting proteins.

or the [^{32}P]-labeled peptide derived from the pGSTag that lacked the receptor subdomain. Interacting proteins were identified in various fractions of the DEAE-enriched brain cytosol. The 250 mM NaCl elution of the 90% ammonium sulfate precipitate contained interacting proteins with apparent molecular weights (kd) of ~ 84, 56, 53, 32 and 30 (Table 1). The same fraction was evaluated for interacting proteins by affinity chromatography using an affinity matrix in which the GST-receptor fusion protein was bound to glutathione-sepharose. This approach identified proteins of apparent molecular weights (kd) of ~ 105, 56, 48, 34 and 32 that were retained by the receptor subdomain-substituted resin but not by the resin containing GST alone (Table 1). Of note is identification of the M_r ~ 56,000 and 32,000 peptides by both gel overlay and the affinity matrix.

TABLE 1
Putative interacting proteins identified by gel overlay and affinity chromatography using a receptor subdomain probe from the third intracellular loop.

gel overlay ($M_r \times 10^{-3}$)	affinity chromatography ($M_r \times 10^{-3}$)
84	105
56	56
53	48
32	34
30	32

DISCUSSION

Two key regulatable steps in the propagation of signals across the cell membrane and within the cell are the exchange of GDP for GTP and the subsequent hydrolysis of GTP by guanine nucleotide binding proteins (Bourne et al., 1991; Hepleer and Gilman, 1992; Neer, 1995). For heterotrimeric G-proteins, binding of agonist to a heptahelical receptor is a trigger for guanine nucleotide exchange. However, the efficiency of this process varies among cell types (Sato et al., 1995; Nanoff et al., 1995), suggesting that additional factors operating within the receptor's microenvironment influence signalling specificity/efficiency for receptors coupled to heterotrimeric guanine nucleotide binding proteins. Such accessory proteins might include kinases and phosphatases that influence receptor trafficking and coupling to G, as well as proteins that together with the receptor participate in the formation of a signal transduction complex on the inner face of the plasma membrane (Neubig, 1994). Such proteins may operate in a cell-type specific manner to fine tune hormone-initiated signals.

One of our current goals is to identify accessory proteins existing in the receptor's microenvironment that perhaps regulate signal propagation or participate in the formation of a signal transduction complex as described in the current report. The strategies described herein may allow us to begin to imagine the receptor's microenvironment and reconstruct a putative signal transduction complex. It is likely that there are multiple G-proteins and effectors around the receptor that compete for the signal intiated by the agonist-activated receptor (Figure 5). In addition, there are other accessory proteins that regulate signal propagation. The latter entities may account for the cell-type specific differences observed for the α_2-AR in terms of signal transfer from R to G (Sato et al., 1995) suggesting the existence of an additional "signal amplifier" or a factor that impedes this event (Sato et al., 1995; Nanoff et al., 1995). Such a factor may be related to the action of recently identified proteins and chemical compounds that directly activate/inhibit G (Scherer et al., 1996; Popova et al., 1994; Strittmatter et al., 1993; Li et al., 1995; Sim et al., 1995; Nishimoto et al., 1993; Okamoto et al., 1995; Higashijima et al., 1988; Mousli et al., 1990). We have recently partially purified such a protein from the neuroblastoma NG108-15 and termed this material a G-protein activator (Sato M., Hildebrandt J.D. and Lanier S.M., unpublished observations). Such molecules may contribute to the amplification of the biological stimuli commonly observed with signalling events involving heterotrimeric G. By reconstructing the receptor's microenvironment and defining the

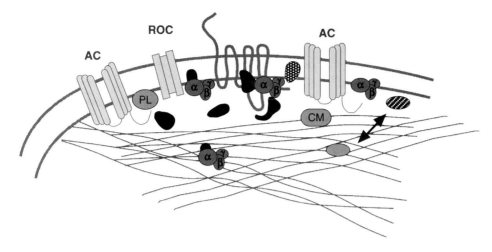

THE RECEPTOR'S MICROENVIRONMENT

FIGURE 5 Imagining the receptor's microenvironment. Refer to text for details. AC: adenylyl cyclase, ROC: receptor operated channel, PL: phospholipase, CM: calmodulin. The arrow indicates signalling molecules that translocate from cytosol to membrane.

precise molecular interactions occuring in the signal transduction cascade, it may be possible to design therapeutic agents that target signalling events distal to receptor activation as opposed to agents that target the receptor's hormone binding site.

ACKNOWLEDGEMENTS

The authors appreciate the contributions of Drs. Emir Duzic, Ryo Kataoka and John D. Hildebrandt to this research effort. This work was supported by grants from the National Institutes of Health NS24821(SML) and the Council for Tobacco Research (#2235)(SML). Dr. Sato's current address is the 1st Department of Internal Medicine, Asahikawa Medical College, Asahikawa Japan.

References

Bourne, H.R., Sanders, D.A. and McCormick, F. (1991). The GTPase superfamily: conserved structure and molecular mechanism. *Nature*, **349**, 117–127.

Duzic, E., Coupry, I., Downing, S. and Lanier, S.M. (1992). Factors determining the specificity of signal transduction by guanine nucleotide-binding protein-coupled receptors. I. Coupling of α_2-adrenergic receptor subtypes to distinct G-proteins. *J. Biol. Chem.*, **267**, 9844–9851.

Coupry, I., Duzic, E. and Lanier, S.M. (1992). Factors determining the specificity of signal transduction by G-protein coupled receptors. II. Preferential coupling of the α_{2C}-adrenergic receptor to the guanine nucleotide binding protein, G_o. *J. Biol. Chem.*, **267**, 9852–9857.

Duzic, E. and Lanier, S.M. (1992). Factors determining the specificity of signal transduction by guanine nucleotide-binding protein-coupled receptors. III. Coupling of α_2-adrenergic receptor subtypes in a cell type-specific manner. *J. Biol. Chem.*, **267**, 24045–24052.

Hamamdzic, D., Duzic, E., Sherlock, J.D. and Lanier, S.M. (1995). Regulation of α_2-adrenergic receptor expression and signaling in pancreatic β-cells. *Amer. J. Physiol.*, **32**, E162–171.

Hepler, J.R. and Gilman, A.G. (1992). G proteins. *Trends Biochem. Sci.*, **17**, 383–387.
Higashijima, T., Uzu, S., Nakajima, T. and Ross, E.M. (1988). Mastoparan, a peptide toxin from wasp venom, mimics receptors by activating GTP-binding regulatory proteins (G proteins). *J. Biol. Chem.*, **263**, 6491–6494.
Lanier, S.M., Downing S., Duzic, E. and Homcy, C.J. (1991). Isolation of rat genomic clones encoding subtypes of the α_2-adrenergic receptor. *J. Biol. Chem.*, **266**, 10470–10478.
Li, S., Okamoto, T., Chun, M., Sargiacomo, M., Casanova, J.E., Hansen S.H., Nishimoto, I. and Lisanti, M.P. (1995) Evidence for a regulated interaction between heterotrimeric G proteins and caveolin. *J. Biol. Chem.*, **270**, 15693–15701.
Mousli, M., Bronner, C., Landry, Y., Bockaert, J. and Rouot, B. (1990). Direct activation of GTP-binding regulatory proteins (G-proteins) by substance P and compound 48/80. *FEBS Lett.*, **259**, 260–262.
Nanoff, C., Mitterauer, T., Roka, F., Hohenegger, M. and Freissmuth, M. (1995). Species differences in A_1 adenosine receptor/G protein coupling: identification of a membrane protein that stabilizes the association of the receptor/G protein complex. *Mol. Pharmacol.*, **48**, 806–817.
Neer, E.J. (1995). Heterotrimeric G proteins: organizers of transmembrane signals. *Cell*, **80**, 249–257.
Neubig, R.R. (1994). Membrane organization in G-protein mechanisms. *FASEB J.*, **8**, 939–946.
Nishimoto, I., Okamoto, T., Matsuura, Y., Takahashi, S., Okamoto, T., Murayama, Y. and Ogata, E. (1993) Alzheimer amyloid protein precursor complexes with brain GTP-binding protein G_o. *Nature*, **362**, 75–79.
Okamoto, T., Takeda, S., Murayama, Y., Ogata, E. and Nishimoto, I. (1995). Ligand-dependent G protein coupling function of amyloid transmembrane precursor. *J. Biol. Chem.*, **270**, 4205–4208.
Popova, J.S., Johnson, G.L. and Rasenick, M.M. (1994). Chimeric $G_{\alpha s}/G_{\alpha i2}$ proteins define domains on $G_{\alpha s}$ that interact with tubulin for β-adrenergic activation of adenylyl cyclase. *J. Biol. Chem.*, **269**, 21748–21754.
Ron, D. and Dressler, H. (1992) pGSTag — a versatile bacterial expression plasmid for enzymatic labeling of recombinant proteins. *BioTechniques*, **13**, 866–869.
Sato, M., Kataoka, R., Dingus, J., Wilcox, M., Hildebrandt, J.D. and Lanier, S.M. (1995). Factors determining specificity of signal transduction by G-protein-coupled receptors. IV. Regulation of signal transfer from receptor to G-protein. *J. Biol. Chem.*, **270**, 15269–15276.
Scherer, P.E., Okamoto, T., Chun, M., Nishimoto, I., Lodish, H.F. and Lisanti, M.P. (1996). Identification, sequence, and expression of caveolin-2 defines a caveolin gene family. *Proc. Natl. Acad. Sci. USA*, **93**, 131–135.
Sim, L.J., Selly, D.E. and Childers, S.R. (1995). *In vitro* autoradiography of receptor-activated G proteins in rat brain by agonist-stimulated guanylyl 5'-[γ-[^{35}S]thio]-triphosphate binding *Proc. Natl. Acad. Sci. USA*, **92**, 7242–7246.
Smith, D.B. and Johnson, K.S. (1988). Single-step purification of polypeptides expressed in *Escherichia coli* as fusions with glutathione S-transferase. *Gene*, **67**, 31–40.
Strittmatter, S.M., Cannon, S.C., Ross, E.M., Higashijima, T. and Fishman, M.C. (1993). GAP-43 augments G protein-coupled receptor transduction in *Xenopus laevis* oocytes. *Proc. Natl. Acad. Sci. USA*, **90**, 5327–5331.
Xu, P., Zot, A.S. and Zot, H.G. (1995). Identification of Acan125 as a myosin-I-binding protein present with myosin-I on cellular organelles of *Acanthamoeba*. *J. Biol. Chem.*, **270**, 25316–25319.

CAVEOLIN, AN INTEGRAL MEMBRANE PROTEIN COMPONENT OF CAVEOLAE MEMBRANES: IMPLICATIONS FOR SIGNAL TRANSDUCTION

MICHAEL P. LISANTI*, SHENGWEN LI, ZHAO-LAN TANG, KENNETH S. SONG, ERIC KÜBLER and PHILIPP E. SCHERER

The Whitehead Institute for Biomedical Research, Nine Cambridge Center, Cambridge, MA 02142-1479

Caveolae are plasma membrane specializations that have been implicated in signal transduction. Caveolin, a 21–24 kD integral membrane protein, is localized to caveolae membranes *in vivo* and serves as a marker protein for this organelle. Purification of caveolin-rich membrane domains from cultured cells and whole tissues reveals several distinct classes of signaling molecules: hetero-trimeric G protein subunits (α and $\beta\gamma$), Src-family tyrosine kinases, PKCα, and certain Ras-related GTPases (Rap family members). Based on these observations, we have proposed that caveolin may function as a scaffolding protein to organize and concentrate caveolin-interacting signaling molecules within caveolae membranes. This "caveolae signaling hypothesis" states that localization of inactive signaling molecules within caveolae membranes could be a compartmental basis for certain transmembrane signaling events. Caveolin is clearly important for the formation of caveolae membranes. Caveolin expression levels directly correlate with the morphological appearance of caveolae during adipocyte differentiation and the disappearance of caveolae during oncogenic transformation of NIH 3T3 cells. In addition, caveolin exists as a high molecular mass oligomer of ~350 kD with ~14–16 caveolin monomers per oligomer. These caveolin homo-oligomers have the capacity to self-associate *in vitro* into larger structures that resemble caveolae. In these structures, caveolin homo-oligomers exhibit side-by-side packing, an indication of how caveolae membranes are constructed. This would provide a platform or scaffold for the recruitment of signaling molecules. In support of this model, caveolin undergoes a number of regulatable modifications that may be critical for this process (i.e., phosphorylation on serine and tyrosine residues and palmitoylation on cysteine residues). Also, caveolin interacts directly with G protein α subunits and can functionally regulate their activity by acting as a GDI (GDP dissociation inhibitor). Another member of the caveolin gene family (designated caveolin-2) acts as a GAP (GTPase activating protein). As both activities place (GAP) or hold (GDI) G_α subunits in the inactive GDP-bound conformation, these two caveolin family members could act in concert to recruit and sequester inactive G-proteins within caveolae membranes — for activation by appropriate receptors.

KEY WORDS: Caveolae; caveolin; G proteins; Src-family tyrosine kinases; membrane microdomains; caveolin gene family

Caveolae are small flask-shaped invaginations located at or near the plasma membrane. They represent a sub-compartment of the plasma membrane. Although they are thought to exist in most cell types, caveolae are most abundant in simple squamous epithelia (endothelial cells and type I pneumocytes), adipocytes, fibroblasts, and smooth muscle cells (reviewed in (Anderson, 1993; Lisanti *et al.*, 1994; Lisanti *et al.*, 1995; Severs, 1988; Travis, 1993)).

Caveolin, a 21–24 kD integral membrane protein, has been identified as a principal component of caveolae membranes *in vivo*. In cells and tissues that express caveolin,

---

Correspondence to: Dr. Michael P. Lisanti. The Whitehead Institute for Biomedical Research, Nine Cambridge Center, Cambridge, MA 02142-1479. Tel. (617) 258-5225; Fax (617) 258-9872; electronic mail: lisanti@wi.mit.edu

caveolin can serve as a marker for the organelle. At steady-state, greater than 90% of caveolin is associated with plasma membrane caveolae (Chang *et al.*, 1994; Rothberg *et al.*, 1992).

However, a lack of caveolin expression should not be taken as an absence of caveolae (Fra *et al.*, 1994), as recent evidence suggests that caveolin may be a regulatory component of caveolae that is not absolutely required to maintain their structure or integrity (Sargiacomo *et al.*, 1993; Smart *et al.*, 1994). In fact, like many other molecules, caveolin may be only the first member of a family of caveolin-related proteins. This could also explain the presence of caveolae in cells and tissues where caveolin is apparently not expressed.

Here, we summarize recent advances in research on caveolin and its relationship to caveolae. In addition, we discuss the possible role of caveolae in cell transformation and transmembrane signalling events.

IDENTIFICATION, CLONING AND REGULATED EXPRESSION OF CAVEOLIN

Primary sequence and domain structure

Cloning of the chicken caveolin cDNA revealed that caveolin is a 178 amino acid protein with a predicted molecular mass of 20.6 kD (Glenney and Soppet, 1992). Although their is no N-terminal hydrophobic signal sequence, caveolin contains a 33-amino acid putative membrane spanning segment (caveolin residues 102–134). Thus, caveolin can roughly be divided into three domains: an N-terminal region (residues 1–101); a membrane spanning region (residues 102-134) and a C-terminal region (residues 135–178) (Figure 1A).

Unlike most membrane spanning segments which are predicted to be α-helical and are in the range of 19–22 amino acids, caveolin's putative membrane attachment site is considerably longer and is predicted to assume a β-conformation (Glenney and Soppet, 1992). This may be related to the recent observation that caveolin can move in and out of plasma membranes and across the ER or Golgi membrane (Smart *et al.*, 1994), behaving both as an integral membrane protein and a soluble cytosolic protein.

Cloning other cDNAs for caveolin has revealed that caveolin is a highly conserved protein molecule (Glenney, 1992; Kurzchalia *et al.*, 1992; Tang *et al.*, 1994). Comparison of the protein sequences for human, canine, murine, and chicken caveolin demonstrates that they are roughly 86% identical from chicken to man (See (Glenney, 1992; Tang *et al.*, 1994) for alignments). In addition, the transmembrane region is most conserved and the C-terminal domain is most variable. Such conservation of the transmembrane region, together with its unusual length and predicted structure, suggest that it may be critical for caveolin's function and novel topology (see below).

Genomic structure

Cloning of the chicken caveolin gene has revealed its intron-exon organization: exon 1, nucleotides 1–30; exon 2, nucleotides 31–191, and exon 3, nucleotides 192–534 (Glenney and Soppet, 1992). Thus the first two exons encode the bulk of the N-terminal domain, while the transmembrane region and C-terminal domain are encoded by the third exon. As caveolin mRNA is a single species (Glenney, 1992; Sargiacomo *et al.*, 1993; Scherer *et al.*, 1994), it is unlikely that the 21 and 24 kD forms of caveolin arise from differential mRNA splicing.

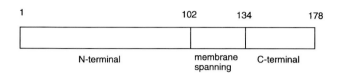

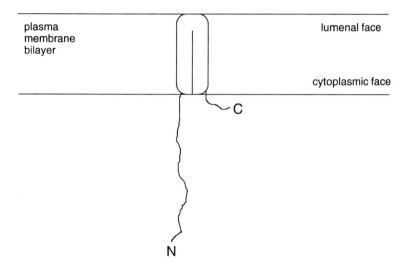

FIGURE 1 A single caveolin molecule. A) Domain organization; B) Membrane topology. Caveolin consists of three domains: an N-terminal region, a membrane spanning segment and a C-terminal domain. Both the N- and C-terminal domains of caveolin have been shown to be entirely cytoplasmic.

Membrane topology

The membrane topology of caveolin has recently been investigated. By analogy with the transferrin receptor, it was initially suggested that caveolin might assume a type II transmembrane orientation — with a cytoplasmic N-terminal domain and an extracellular C-terminal domain (Glenney, 1992; Kurzchalia *et al.*, 1992). However, analysis of the caveolin cDNA reveals an unusual 33-amino acid hydrophobic membrane spanning segment which could form a hair-pin like structure within the membrane, allowing both the N-terminal and C-terminal domains of caveolin to remain entirely cytoplasmic (Glenney, 1992; Kurzchalia *et al.*, 1992) (Figure 1B). The cytoplasmic orientation of both the

N-terminal and C-terminal domains of caveolin has now been confirmed by epitope-tagging, antibody accessibility and immuno-gold localization studies (Dupree et al., 1993; Scherer et al., 1995; Dietzen et al., 1995).

Expression

Lung and muscle are tissues rich in caveolin mRNA and protein; lower levels of caveolin expression are detected in brain, while no detectable caveolin mRNA is found in spleen, kidney, liver and testis (Glenney, 1992; Lisanti et al., 1994; Scherer et al., 1994). However, the most abundant source of caveolin is white adipose tissue (Scherer et al., 1994). This is consistent with previous morphological studies suggesting that differentiated 3T3-Ll adipocytes are a rich source of caveolae (Fan et al., 1983). During adipocyte differentiation, caveolin mRNA and protein are induced over 20-fold (Scherer et al., 1994) while there is a concomitant nine-fold increase in the number of caveolae-like structures (Fan et al., 1983).

The abundance of caveolin in insulin-sensitive tissues (adipose and muscle) suggests that it might participate in insulin-mediated signal transduction events. In support of this notion, the ligand-bound insulin receptor is localized within adipocyte caveolae (Goldberg et al., 1987) and in response to insulin stimulation an uncharacterized 32 kD phosphoprotein rapidly associates with caveolin in adipocytes (Scherer et al., 1994).

Serine phosphorylation

Caveolin is constitutively phosphorylated on serine residues (Glenney, 1989; Sargiacomo et al., 1994; Sargiacomo et al., 1993). Interestingly, only the faster migrating 21 kD species of caveolin is phosphorylated *in vivo*, while both forms are capable of undergoing serine phosphorylation *in vitro* (Scherer et al., 1994). These observations point to a functional difference between the 24 kD and 21 kD species of caveolin and this functional difference appears to be recognized by a specific serine kinase *in vivo*.

The identity of this serine kinase remains unknown. Caveolin contains consensus sites for phosphorylation by serine-threonine kinases — protein kinase C [Ser-37] and casein kinase II [Ser 88] (Sargiacomo et al., 1994; Tang et al., 1994). In addition, both PKC α and CK II are concentrated in purified caveolin-rich membrane domains (Lisanti et al., 1994; Sargiacomo et al., 1994). Serine phosphoration is apparently important for regulating the function of caveolae as i) PKC activators cause caveolae to flatten out and prevents uptake of folate via caveolar potocytosis (Smart et al., 1993) and ii) a serine phosphatase inhibitor (okadaic acid) dramatically alters the subcellular distribution of caveolae (Parton et al., 1994).

Relationship with cell transformation.

Caveolin was first identified as a major v-Src substrate in RSV-transformed chick embryo fibroblasts (Glenney, 1989; Glenney and Zokas, 1989). As both cell transformation and tyrosine phosphorylation of caveolin are dependent on membrane attachement of v-Src (Glenney, 1989), it has been suggested that caveolin may represent a critical substrate for cellular transformation.

In support of this view, we have recently observed that both caveolin expression and caveolae are lost during cell transformation by activated oncogenes other than v-Src [v-abl, bcr-abl, middle T antigen and activated ras] (Koleske et al., 1995). These results support the hypothesis that caveolin may represent a candidate tumor suppressor protein. Indeed, Krev-1 — a ras-related transformation suppressor protein — is concentrated in purified caveolin-rich membrane domains (Lisanti *et al.*, 1994) and purified caveolae (Chang *et al.*, 1994).

USE OF CAVEOLIN AS A MARKER FOR THE PURIFICATION OF CAVEOLAE AND THE IDENTIFICATION OF CAVEOLIN-ASSOCIATED PROTEINS

Caveolae purification and the 'caveolae signalling hypothesis'

As caveolin is a principal component of caveolae membranes, we and others have used caveolin as a biochemical marker for the purification of caveolae from cultured cells and whole tissue. As the exact function of caveolae is unknown, identification of the components of caveolin-enriched membrane fractions could provide a systematic basis to experimentally examine caveolae functioning.

These purified caveolin-enriched membrane domains quantitatively exclude markers for other subcellular compartements such as plasma membrane (non-caveolar), Golgi, lysosomes, mitochondria, and ER, suggesting that they are relatively pure (Chang *et al.*, 1994; Lisanti *et al.*, 1994; Lisanti *et al.*, 1995; Sargiacomo *et al.*, 1993; Scherer *et al.*, 1994; Smart *et al.*, 1994). In addition, greater than 90% of these vesicular structures and curved membrane fragments are immuno-labeled with anti-caveolin IgG (Chang *et al.*, 1994; Lisanti *et al.*, 1994).

Microsequencing and immunoblotting of caveolin-enriched domains reveals several distinct classes of molecules (Chang *et al.*, 1994; Lisanti *et al.*, 1994). These include: scavenger receptors for oxidized LDL and AGEs, namely CD 36 and RAGE; components of the actin based cytoskeleton; hetero-trimeric G proteins (α and $\beta\gamma$ subunits); Src-like kinases; Rap GTPases; and other cytoplasmic signalling molecules. These results have recently been confirmed using a different scheme to purify caveolae from endothelial cell plasma membranes (Schnitzer *et al.*, 1995).

Many independent biochemical and morphological observations also support the above observations: caveolae transcytose atherogenic forms of LDL and participate in scavenger endocytosis of albumin (Snelting-Havinga *et al.*, 1989); AGE's and their receptor (RAGE) have been localized to caveolae by immuno-gold labeling (Schmidt *et al.*, 1994); caveolae membranes can be labeled *in vivo* with an actin-binding protein, namely myosin sub-fragment I (Izumi *et al.*, 1988); G-protein coupled receptors undergo ligand induced translocation to caveolae (Lisanti, 1994; Raposo and Benedetti, 1994) or are constituitively localized within caveolae (Chun *et al.*, 1994), G protein modifying bacterial toxins (cholera toxin) enter cells via caveolae (Montesano *et al.*, 1982; Parton, 1994).

Recent mutational analysis indicates that a signal for dual acylation (MGCXXS/C; termed an "SH-4 domain") can act as a caveolar localization signal for Src-like kinases and perhaps hetero-trimeric G-protein α subunits (see below) (Chang *et al.*, 1994; Lisanti *et al.*, 1994; Shenoy-Scaria *et al.*, 1994).

Based on these biochemical and morphological observations, we and others have recently proposed the hypothesis that caveolae may participate in a subset of transmembrane signalling events (Chang *et al.*, 1994; Lisanti *et al.*, 1994; Shenoy-Scaria *et al.*, 1994). This 'caveolae signaling hypothesis' states that "compartmentalization of certain

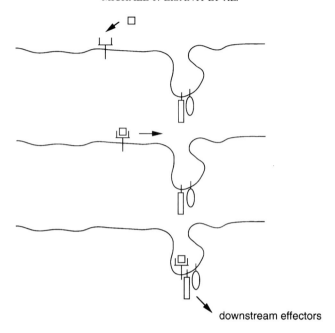

FIGURE 2 The 'caveolae signaling hypothesis'. G-protein coupled receptors (such as the β-adrenergic receptor and the muscarinic acetylcholine receptor) undergo ligand-induced caveolar localization. After entering a caveolae, these activated receptors could then select a given lipid-modified effector molecule, including heterotrimeric G-proteins and Src-like kinases, to mediate downstream signalling events. Redrawn from (Lisanti et al., 1995).

signalling molecules within caveolae could allow efficient and rapid coupling of activated receptors to more than one effector system" (Lisanti et al., 1994; Lisanti et al., 1995) (Figure 2).

Thus, we might envision signalling through caveolae as a two-step process: first, conformational changes could serve as agonist-induced targeting signals to localize activated receptors within caveolae and second, activated receptors could then efficiently select one or more of the appropriate effectors systems for signal transduction (Lisanti et al., 1994; Lisanti et al., 1995). This hypothesis is testable and allows for cross-talk between different effector systems, while providing for specificity and the dynamic nature of signaling events.

In further support of this hypothesis, we have recently observed that: i) a G-protein coupled receptor — containing its bound ligand — forms a tight complex with caveolin at the level of the plasma membrane (Chun et al., 1994); and ii) caveolin rapidly associates with a 32 kD phosphoprotein in response to insulin stimulation (Scherer et al., 1994). The ligand-bound insulin receptor has been previously localized to caveolae (Goldberg et al., 1987).

Dual acylation and caveolar localization

Recently, an N-terminal consesus sequence for dual acylation of cytoplasmically oriented signalling molecules was identified. This sequence, MGCX2-3[S/C], termed an SH4 domain (Src-homology 4 domain; (Resh, 1994)) is common to the cytoplasmic N-terminus of multiple Src-like kinases, heterotrimeric G-protein α subunits and the transmembrane glycoprotein CD 36. Given the common localization of these components within caveolin-rich domains, dual acylation by myristate and palmitate could serve as a regulatable signal for caveolar localization. Unlike myristoylation, palmitoylation is a dynamic reversible modification that can be regulated by palmitoyl-thioesterases.

The above hypothesis is supported by the observations that 1) N-terminal myristoylation of v-Src is required for v-Src mediated tyrosine phosphorylation of caveolin *in vivo* (Glenney, 1989); and 2) dual acylation of Src like kinases facilitates their co-immunoprecipitation with GPI-linked proteins (Shenoy-Scaria *et al.*, 1993). This hypothesis has also recently been confirmed by studies in which myristoylation or pamitoylation of Src-family tyrosine kinases is prevented by site-directed mutagenesis (Robbins *et al.*, 1995; Shenoy-Scaria *et al.*, 1994)

CAVEOLIN: STRUCTURE-FUNCTION AND DOMAIN MAPPING

These studies are summarized schematically in Figure 3.

Caveolin isoforms

Two isoforms of caveolin are known to exist: a slower migrating 24 kD form and a faster migrating 21 kD form. We have designated these as α-caveolin and β-caveolin, respectively (Scherer *et al.*, 1995). As we have shown that only the β-isoform undergoes serine phosphorylation *in vivo*, this points to a functional difference between these isoforms that is recognized by a specific serine kinase *in vivo* (Scherer *et al.*, 1994).

Recently, we have determined that α- and β-caveolin differ in their N-terminal protein sequence. α-caveolin contains residues 1–178 and β-caveolin contains residues 32-178 (Scherer *et al.*, 1995). Both derive from a single cDNA. We have shown that these two isoforms are generated by alternate initiation of translation using two alternate start sites: methionine at position 1 and methionine at position 32 (Scherer *et al.*, 1995). This explains their apparent 3 kD difference in migration in SDS-PAGE gels.

As both isoforms form oligomers of caveolin (see below) and both isoforms are targeted to caveolae, caveolin residues 1–31 are apparently not required for these caveolin functions. However, immunofluorescent localization of these isoforms using isoform-specific antibody probes reveals that they may assume a distinct but overlapping subcellular distribution (Scherer *et al.*, 1995). Based on these observations, we have suggested that at least two distinct sub-populations of caveolae may exist.

In addition, both isoforms are targeted to caveolae-like membrane fractions when exogenously expressed in a caveolin-negative cell line (Scherer *et al.*, 1995). This is consistent with previous results suggesting that caveolin is not necessarily required to maintain the structure or integrity of caveolae (Sargiacomo *et al.*, 1993; Smart *et al.*, 1994; Zurzolo *et al.*, 1994).

Epitope mapping of isoform specific mAb probes

While studying caveolin isoforms, we fortuitously identified an α-isoform specific mAb probe (2234). We have shown that mAb 2234 recognizes an epitope within caveolin residues 1–21, while mAb 2297 recognizes an epitope within caveolin residues 61–71 (Scherer *et al.*, 1995). As expected mAb 2297 recognizes both α- and β-caveolin (Scherer *et al.*, 1995).

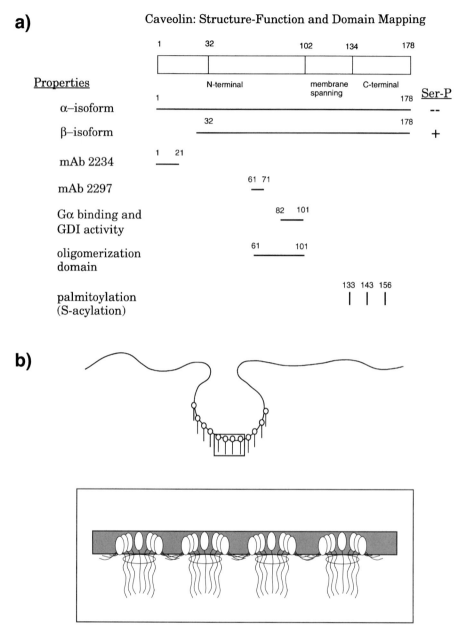

FIGURE 3 Caveolin: Structure-function and domain mapping. A) The overall domain organization of caveolin is shown above. Functional properties of caveolin are listed at Left and caveolin residues participating in these functions are as indicated. These properties include the structure of caveolin isoforms, mAb epitopes, G_α binding and GDI activities, caveolin oligomerization domain and palmitoylation sites. Note that differential serine phosphorylation of caveolin isoforms is indicated at Right. B) Potential organization of caveolin homo-oligomers within caveolae membranes. Each caveolin homo-oligomer (shown in cross-section) should contain ~14 caveolin monomers. These caveolin homo-oligomers may undergo side-by-side packing to form caveolae and to act as a scaffold to recruit signaling molecules. The ring surrounding the N-terminal domain of each caveolin homo-oligomers indicates that membrane proximal residues 61–101 of caveolin are responsible for homo-oligomer formation (Sargiacomo et al., 1995).

Identification and epitope mapping of these mAbs provides molecular probes to structurally and functionally distinguish between the two isoforms of caveolin.

G_α subunit binding and GDI activity

We have recently presented several independent lines of evidence suggesting that caveolin interacts with G_α subunits in a regulated fashion and that caveolin functions as a GDI for hetero-trimeric G proteins (Li *et al.*, 1995).

This interaction is highly specific and regulated by the activation state of the G_α subunit (Li *et al.*, 1995). We have shown that only *inactive* G_α subunits co-fractionate with caveolin, that *inactive* G_α subunits preferentially interact directly with recombinant caveolin, and that caveolin-derived polypeptides functionally *suppress* the basal activity of purified hetero-trimeric G proteins — apparently by inhibiting GDP/GTP exchange (Li *et al.*, 1995). We have mapped this functional GDI activity to residues 82–101 of the caveolin molecule; this caveolin region bears striking homology to a region of rab GDI that is conserved from yeast to man. These results are consistent with the hypothesis that caveolae may function as membrane-bound sites to store certain *inactive* cytoplasmically oriented signaling molecules, as we have suggested earlier (Koleske *et al.*, 1995; Lisanti *et al.*, 1994).

Caveolin subunit structure and oligomerization domain

We and others have suggested that caveolin may function as a scaffolding protein to organize and concentrate signaling molecules within caveolae membranes (Lisanti *et al.*, 1994; Neer, 1995; Neubig, 1994). In support of this view, we have determined that inactive G_α subunits interact with caveolin in a 1:1 stoichiometry and that caveolin binding functionally suppresses the activity of trimeric G proteins — clamping them in an inactive conformation (Li *et al.*, 1995). Thus, knowledge of the subunit structure of caveolin is important for understanding how caveolin might function to concentrate G_α subunits or other caveolin-interacting signaling molecules within caveolae membranes.

Recently, we and others have determined that caveolin exists as a homo-oligomer of ~14 individual caveolin monomers (Monier *et al.*, 1995; Sargiacomo *et al.*, 1995). In addition, we have defined a minimal 41-amino acid region of the caveolin N-terminal cytoplasmic domain that can confer oligomerization of the same relative stoichiometry in the context of a GST-fusion protein. We have shown that GST does not contribute to this homo-oligomerization event (Sargiacomo *et al.*, 1995). Both α- and β-caveolin form homo- and hetero-dimers within these oligomers of caveolin and we have also shown that caveolin oligomers exist within intact plasma membranes (Sargiacomo *et al.*, 1995).

As caveolin also interacts with G_α subunits, self-oligomerization of caveolin could provide a means for concentrating inactive trimeric G-proteins and other caveolin-interacting proteins within caveolae. In this regard, we have suggested that i) caveolin oligomerization may provide a general mechanism for concentrating caveolin-interacting molecules within caveolae and ii) it may be useful to think of caveolin homo-oligomers as 'fishing lures' with multiple 'hooks' or attachment sites for caveolin-interacting signaling molecules (Sargiacomo *et al.*, 1995).

What could be the structural basis for caveolin homo-oligomerization? We have suggested that caveolin homo-oligomerization may be based on a known structural motif — called the α/β-coiled fold (See Sargiacomo *et al.*, 1995 for a more detailed discussion of this point; (Sargiacomo *et al.*, 1995)). Confirmation of this assertion will require determination of the crystal structure of caveolin.

Palmitoylation and membrane attachment

Caveolin contains three conserved cysteine residues that are potential sites for fatty-acylation by palmitate. Lublin and colleagues have recently shown that all three cysteines (caveolin residues 133, 143, and 156) undergo palmitoylation, but membrane-proximal cysteine 133 is the major site of palmitoylation (Dietzen *et al.*, 1995). As palmitoylation occurs only on cytoplasmically oriented cysteine residues, these results independently demonstrate that the C-terminal domain of caveolin is cytoplasmically oriented.

In addition, site-directed mutagenesis of these residues eliminates palmitate incorporation, however this does not prevent caveolar localization or membrane attachment (Dietzen *et al.*, 1995). Thus, palmitoylation of the C-terminal domain is not required for caveolar localization.

CAVEOLIN DIVERSITY: IDENTIFICATION OF A NOVEL CAVEOLIN-RELATED PROTEIN

In cells and tissues where caveolin is expressed, caveolin can serve as a marker for the organelle (Rothberg *et al.*, 1992). At steady-state, greater than 90% of caveolin is associated with plasma membrane caveolae (Chang *et al.*, 1994; Smart *et al.*, 1994). Like many other molecules, however, caveolin may be only the first member of a family of caveolin-related proteins.

Recently, we have obtained evidence for the existence of a novel caveolin-related protein. A novel ~20 kD caveolin-related protein, caveolin-2, was identified through micro-sequencing of adipocyte-derived caveolin-enriched membranes; caveolin was re-termed caveolin-1 (Scherer *et al.*, 1996).

Caveolins 1 and 2 are similar in many respects. Human caveolin-2 is ~38% identical and 58% similar to human caveolin-1 (Figure 4A). The most conserved region is a stretch of eight identical amino acids (FEDVIAEP) within the N-terminal domains of caveolin-1 and caveolin-2. This may represent a "signature sequence" that is characteristic of members of the caveolin gene family. Both caveolin-1 and caveolin-2 contain a 33-amino acid membrane spanning segment and a hydrophilic 43–44-amino acid C-terminal domain. However, the N-terminal domain of caveolin-2 is 26 amino acids shorter than caveolin-1α. Thus, caveolin-1β (147 aa) is approximately the same length as caveolin-2. In addition, mRNA's for both caveolin-1 and caveolin-2 are most abundantly expressed in white adipose tissue and are induced during adipoctye differentiation.

Is the co-expression of caveolin-1 and caveolin-2 within a single cell functionally redundant or do they serve distinct but complementary roles? As caveolin-2 co-localizes with caveolin-1 by cell fractionation and immuno-localization studies, caveolin-2 must also localize to caveolae. However, some evidence suggests that caveolin-1 and caveolin-2 are functionally distinct. Caveolin-1 may function as a GDI (GDP dissociation inhibitor) for hetero-trimeric G-proteins. A peptide derived from caveolin-1 functionally interacts with purified trimeric G-proteins and suppresses basal GTPase activity (Li *et al.*, 1995). In contrast, a peptide derived from the corresponding sequence of caveolin-2 functionally activates the GTPase activity of purified trimeric G-proteins without affecting GDP/GTP

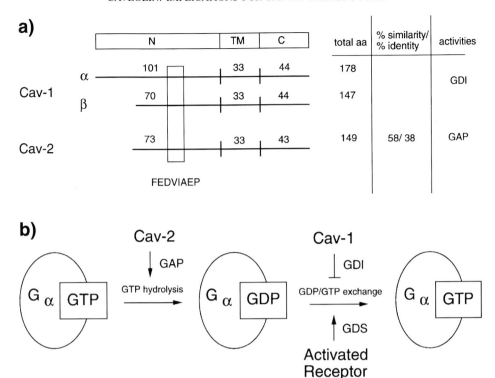

FIGURE 4 Schematic diagram summarizing known caveolin family members. A) The overall structure of caveolin-1 (α- and β-isoforms) and caveolin-2 is shown. All three protein products contain the invariant sequence FEDVIAEP within their hydrophilic N-terminal domains. Note that caveolin-1β (147 aa) is approximately the same length as caveolin-2 (149 aa) and caveolin-3 (151 aa); % similarity and identity of caveolins 2 and 3 with caveolin-1 are shown. GDI and GAP activities are also noted. GDI, GDP-dissociation inhibitor; GAP, GTPase activating protein. B) We have shown that a caveolin-2 derived polypeptide acts as a GAP to actively place the G protein in the inactive GDP-bound state, while the corresponding caveolin–1 derived polypeptide acts as a GDI to hold the G protein in the inactive conformation by preventing GDP-GTP exchange (Li et al., 1995; Scherer et al., 1995). As both activities place or hold G α subunits in the inactive conformation, this suggests that caveolin-1 and caveolin-2 might act in concert to sequentially recruit and sequester inactive G proteins within caveolae membranes.

exchange; caveolin 2 functions as a GAP (GTPase activating protein) (Scherer et al., 1995).

Both the GAP and GDI activities serve to place or hold G proteins in the inactive GDP-liganded conformation, suggesting that caveolin-1 and caveolin-2 might act in concert to sequentially recruit and sequester inactive G proteins within caveolae membranes (Figure 4B). Specifically, i) caveolin-2 could function first as a GAP to place activated G proteins in the inactive conformation and recruit them to the caveolae membrane and ii) caveolin-1 could then function as a GDI to hold them in an inactive conformation within caveolae membranes. This would provide a two-step mechanism for concentrating inactive G proteins within caveolae for presentation to activated G-protein coupled receptors. Consistent with this idea, ligand-bound endothelin receptors co-localize with caveolin-1 within the plasma membrane (Chun et al., 1994). The mechanism by which G proteins cycle on and off the plasma membrane remains largely unknown (Wedegaertner et al.,

1995). No other integral membrane protein co-factors have been previously identified that could facilitate or regulate this process. In this regard, our current findings could provide a new framework for understanding these molecular signaling events in the context of caveolae.

CONCLUDING REMARKS

In conclusion, the discovery of caveolin and a family of caveolin-like molecules suggests that caveolin family members may act as scaffolding proteins to recruit and sequester signaling molecules within discrete microdomains of the plasma membranes in a variety of cell types. This diversity may also provide for the generation of different subpopulations of caveolae involved in related but distinct transmembrane signaling pathways.

ACKNOWLEDGEMENTS

We wish to thank Harvey F. Lodish for his enthusiasm and encouragement. This work was funded by a grant from the W.M. Keck Foundation to the Whitehead Fellows program (to M.P.L.), an NIH FIRST Award GM-50443 (to M.P.L.) and a grant from the Elsa U. Pardee Foundation (M.P.L.). E.K. was supported by a fellowship from the Swiss National Science Foundation. P.E.S was supported by a Swiss National Science Foundation fellowship and NIH grants (GM-49516/ DK-47618) to H. Lodish.

References

Anderson, R.G.W. (1993). Plasmalemmal caveolae and GPI-anchored membrane proteins. *Current Opinion in Cell Biology*, **5**, 647–652.

Chang, W.-J., Ying, Y., Rothberg, K.G., Hooper, N.M., Turner, A.J., Gambiel, H., De Gunzburg, J., Mumby, S.M., Gilman, A.G. and Anderson, R.G.W. (1994a). Purification and characterization of smooth muscle cell caveolae. *J. Cell Biol.*, **126**, 127–138.

Chang, W.J., Ying, Y., Rothberg, K., Hooper, N., Turner, A., Gambliel, H., De Gunzburg, J., Mumby, S., Gilman, A. and Anderson, R.G.W. (1994b). Purification and characterization of smooth muscle cell caveolae. *J. Cell Biol.*, **126**, 127–138.

Chun, M., Liyanage, U., Lisanti, M.P. and Lodish, H.F. (1994). Signal transduction of a G-protein coupled receptor in caveolae: Colocalization of endothelin and its receptor with caveolin. *Proc. Natl. Acad. Sci., USA*, **91**, 11728–11732.

Dietzen, D.J., Hastings, W.R. and Lublin, D.M. (1995). Caveolin is palmitoylated on multiple cysteine residues: Palmitoylation is not necessary for localization of caveolin to caveolae. *J. Biol. Chem.*, **270**, 6838–6842.

Dupree, P., Parton, R.G., Raposo, G., Kurzchalia, T.V. and Simons, K. (1993). Caveolae and sorting of the trans-Golgi network of epithelial cells. *EMBO J.*, **12**, 1597–1605.

Fan, J.Y., Carpentier, J.-L., van Obberghen, E., Grunfeld, C., Gorden, P. and Orci, L. (1983). Morphological changes of the 3T3-L1 fibroblast plasma membrane upon differentiation to the adipocyte form. *J. Cell Sci.*, **61**, 219–230.

Fra, A.M., Williamson, E., Simons, K. and Parton, R.G. (1994). Detergent-insoluble glycolipid microdomains in lymphocytes in the absence of caveolae. *J. Biol. Chem.*, **269**, 30745–30748.

Glenney, J.R. (1989). Tyrosine phosphorylation of a 22 kD protein is correlated with transformation with Rous sarcoma virus. *J. Biol. Chem.*, **264**, 20163–20166.

Glenney, J.R. (1992). The sequence of human caveolin reveals identity with VIP 21, a component of transport vesicles. *FEBS Lett.*, **314**, 45–48.

Glenney, J.R. and Soppet, D. (1992). Sequence and expression of caveolin, a protein component of caveolae plasma membrane domains phosphorylated on tyrosine in RSV-transformed fibroblasts. *Proc. Natl. Acad. Sci., USA*, **89**, 10517–10521.

Glenney, J.R. and Zokas, L. (1989). Novel tyrosine kinase substrates from Rous sarcoma virus transformed cells are present in the membrane cytoskeleton. *J. Cell Biol.*, **108**, 2401–2408.

Goldberg, R.I., Smith, R.M. and Jarett, L. (1987). Insulin and α-2 macroglobulin undergo endocytosis by different mechanisms in rat adipocytes. *J. Cellular Phys.*, **133**, 203–212.

Izumi, T., Shibata, Y. and Yamamoto, T. (1988). Striped structures on the cytoplasmic surface membranes of endothelial vesicles of the rat aorta revealed by quick-freeze, deep-etching replicas. *Anat. Rec.*, **220**, 225–232.

Koleske, A.J., Baltimore, D. and Lisanti, M.P. (1995). Reduction of caveolin and caveolae in oncogenically transformed cells. *Proc. Natl. Acad. Sci., USA*, **92**, 1381–1385.

Kurzchalia, T., Dupree, P., Parton, R.G., Kellner, R., Virta, H., Lehnert, M. and Simons, K. (1992). VIP 21, A 21-kD membrane protein is an integral component of trans-Golgi-network-derived transport vesicles. *J. Cell Biol.*, **118**, 1003–1014.

Li, S., Okamoto, T., Chun, M., Sargiacomo, M., Casanova, J.E., Hansen, S.H., Nishimoto, I. and Lisanti, M.P. (1995). Evidence for a regulated interaction of hetero-trimeric G proteins with caveolin. *J. Biol. Chem.*, **270**, 15693–15701.

Lisanti, M.P. (1994). β adrenergic receptors in caveolae? *Trends in Cell Biology*, **4**, 354.

Lisanti, M.P., Scherer, P., Tang, Z.-L. and Sargiacomo, M. (1994a). Caveolae, caveolin and caveolin-rich membrane domains: A signalling hypothesis. *Trends In Cell Biology*, **4**, 231–235.

Lisanti, M.P., Scherer, P.E., Vidugiriene, J., Tang, Z.-L., Hermanoski-Vosatka, A., Tu, Y.-H., Cook, R.F. and Sargiacomo, M. (1994b). Characterization of caveolin-rich membrane domains isolated from an endothelial-rich source: Implications for human disease. *J. Cell Biol.*, **126**, 111–126.

Lisanti, M.P., Scherer, P.E., Tang, Z.-L., Kubler, E., Koleske, A.J. and Sargiacomo, M.S. (1995a). Caveolae and human disease: Functional roles in transcytosis, potocytosis, signalling and cell polarity. Seminars in Developmental Biology **6**, 47–58.

Lisanti, M.P., Tang, Z.-T., Scherer, P. and Sargiacomo, M. (1995b). Caveolae purification and GPI-linked protein sorting in polarized epithelia. *Meth. Enzymol.*, **250**, 655–668.

Monier, S., Parton, R.G., Vogel, F., Behlke, J., Henske, A. and Kurzchalia, T. (1995) VIP21-caveolin, a membrane protein constituent of the caveolar coat, oligomerizes *in vivo* and *in vitro*. *Mol. Biol. Cell*, **6**, 911–927.

Montesano, R., Roth, J., Robert, A. and Orci, L. (1982). Non-coated membrane invaginations are involved in binding and internalization of cholera and tetanus toxins. *Nature* (Lond.) **296**, 651–653.

Neer, E.J. (1995). Hetero-trimeric G proteins: Organizers of transmembrane signals. *Cell*, **80**, 249–257.

Neubig, R.R. (1994). Membrane organization in G-protein mechanisms. *FASEB J.*, **8**, 939–946.

Parton, R.G. (1994). Ultrastructural localization of gangliosides: GM1 is concentrated in caveolae. *J. Histochem. Cytochem.*, **42**, 155–166.

Parton, R.G., Joggerst, B. and Simons, K. (1994). Regulated internalization of caveolae. *J. Cell Biol.*, **127**, 1199–1215.

Raposo, G. and Benedetti, E.L. (1994). Are β-adrenergic receptors internalized via caveolae or coated pits? *Trends in Cell Biology*, **4**, 418.

Resh, M.D. (1994). Myristoylation and palmitoylation of Src family members: The fats of the matter. *Cell*, **76**, 411–413.

Robbins, S.M., Quintrell, N.A. and Bishop, M.J. (1995). Myristoylation and differential palmitoylation of the HCK protein tyrosine kinases govern their attachment to membranes and association with caveolae. *Mol. Cell. Biol.*, **15**, 3507–3515.

Rothberg, K.G., Heuser, J.E., Donzell, W.C., Ying, Y., Glenney, J.R. and Anderson, R.G.W. (1992). Caveolin, a protein component of caveolae membrane coats. *Cell*, **68**, 673–682.

Sargiacomo, M., Sudol, M., Tang, Z.-L. and Lisanti, M.P. (1993). Signal transducing molecules and GPI-linked proteins form a caveolin-rich insoluble complex in MDCK cells. *J. Cell Biol.*, **122**, 789–807.

Sargiacomo, M., Scherer, P.E., Tang, Z.-L., Casanova, J.E. and Lisanti, M.P. (1994). *In vitro* phosphorylation of caveolin-rich membrane domains: Identification of an associated serine kinase activity as a casein kinase II-like enzyme. *Oncogene*, **9**, 2589–2595.

Sargiacomo, M., Scherer, P.E., Tang, Z.-L., Kubler, E., Song, K.S., Sanders, M.C. and Lisanti, M.P. (1995). Oligomeric structure of caveolin: Implications for caveolae membrane organization. *Proc. Natl. Acad. Sci., USA*, **92**, 9407–9411.

Scherer, P.E., Lisanti, M.P., Baldini, G., Sargiacomo, M., Corley-Mastick, C. and Lodish, H.F. (1994). Induction of caveolin during adipogenesis and association of GLUT4 with caveolin-rich vesicles. *J. Cell Biol.*, **127**, 1233–1243.

Scherer, P.E., Okamoto, T., Chun, M., Nishimoto, I., Lodish, H.F. and Lisanti, M.P. (1996). Identification, sequence and expression of caveolin-2 defines a caveolin gene family. *Proc. Natl. Acad. Sci., USA*, **93**, 131–135.

Scherer, P.E., Tang, Z.-L., Chun, M.C., Sargiacomo, M., Lodish, H.F. and Lisanti, M.P. (1995). Caveolin isoforms differ in their N-terminal protein sequence and subcellular distribution: Identification and epitope mapping of an isoform-specific monoclonal antibody probe. *J. Biol. Chem.*, **270**, 16395–16401.

Schmidt, A., Hasu, M., Popov, D., Zhang, J., Chen, J., Yan, S., Brett, J., Cao, R., Kuwabara, K., Costache, G., Simionescu, N., Simionescu, M. and Stern, D. (1994). Receptor for advanced glycosylation endproducts (RAGE) has a central role in vessel wall interactions and gene activation in response to circulating AGE proteins. *Proc. Natl. Acad. Sci., USA*, **91**, 8807–8811.

Schnitzer, J., McIntosh, D., Dvorak, A.M., Liu, J. and Oh, P. (1995). Separation of caveolae from associated microdomains of GPI-anchored proteins. *Science*, **269**, 1435–1439.

Severs, N.J. (1988). Caveolae: Static inpocketings of the plasma membrane, dynamic vesicles or plain artifact. *J. Cell Sci.*, **90**, 341–348.

Shenoy-Scaria, A., Gauen, L., Kwong, J., Shaw, A. and Lublin, D. (1993). Palmitoylation of an N-terminal cysteine motif of protein tyrosine kinases p 56 lck and p 59 tyn mediates interaction with GPI-anchored proteins. *Mol. Cell Biol.*, **13**, 6385–6392.

Shenoy-Scaria, A.M., Dietzen, D.J., Kwong, J., Link, D.C. and Lublin, D.M. (1994). Cysteine-3 of Src family tyrosine kinases determines palmitoylation and localization in caveolae. *Journal of Cell Biology*, **126**, 353–363.

Smart, E.J., Foster, D., Ying, Y.-S., Kamen, B.A. and Anderson, R.G.W. (1993). Protein kinase C activators inhibit receptor-mediated potocytosis by preventing internalization of caveolae. *J. Cell Biol.*, **124**, 307–313.

Smart, E., Ying, Y.-S., Conrad, P. and Anderson, R.G.W. (1994). Caveolin moves from caveolae to the Golgi apparatus in response to cholesterol oxidation. *J. Cell Biol.*, **127**, 1185–1197.

Snelting-Havinga, I., Mommaas, M., Van Hinsbergh, V., Daha, M., Daems, W. and Vermeer, B. (1989). Immunoelectron microscopic visualization of the transcytosis of low density lipoproteins in perfused rat arteries. *Eur. J. Cell Biol.*, **1989**, 27–36.

Tang, Z.-L., Scherer, P.E. and Lisanti, M.P. (1994). The primary sequence of murine caveolin reveals a conserved consensus site for phosphorylation by protein kinase C. *Gene*, **147**, 299–300.

Travis, J. (1993). Cell biologists explore 'tiny caves'. *Science*, **262**, 1208–1209.

Wedegaertner, P.B., Wilson, P.T. and Bourne, H.R. (1995). Lipid modification of trimeric G proteins. *J. Biol. Chem.*, **270**, 503–506.

Zurzolo, C., van't Hof, W., van Meer, G. and Rodriguez-Boulan, E. (1994). VIP21/caveolin, glycosphingolipid clusters and the sorting of GPI-anchored proteins in epithelial cells. *EMBO J.*, **13**, 42–53.

Note Added in Proof

The following is a partial list of caveolae-related articles that have appeared since the submission of this review. These articles directly support many of the ideas discussed within this review.

Fra, A.M., Masserini, M., Palestini, P., Sonnino, S. and Simons, K. (1995). A photo-reactive derivative of ganglioside GM1 specifically cross-links VIP21-caveolin on the cell surface. *FEBS Lett.*, **375**, 11–14.

Li, S., Seitz, R. and Lisanti, M.P. (1996). Phosphorylation of caveolin by Src tyrosine kinases: The α-isoform of caveolin is selectively phosphorylated by v-Src *in vivo*. *J. Biol. Chem.*, **271**, 3863–3868.

Li, S., Song, K.S. and Lisanti, M.P. (1996). Expression and characterization of recombinant caveolin: Purification by poly-histidine tagging and cholesterol-dependent incorporation into defined lipid membranes. *J. Biol. Chem.*, **271**, 568–573.

Liu, P., Ying, Y., Ko, Y.-G. and Anderson, R.G.W. (1996). Localization of the PDGF-stimulated phosphorylation cascade to caveolae. *J. Biol. Chem.*, **271**, 10299–10303.

Parolini, I., Sargiacomo, M., Lisanti, M.P. and Peschle, C. (1996). Signal transduction and GPI-linked proteins (Lyn, Lck, CD4, CD45, G proteins, CD 55) selectively localize in Triton-insoluble plasma membrane domains of human leukemic cell lines and normal granulocytes. *Blood*, **87**, 3783–3794.

Shaul, P.W., Smart, E.J., Robinson, L.J., German, Z., Yuhanna, I.S., Ying, Y., Anderson, R.G.W. and Michel, T. (1996). Acylation targets endothelial nitric-oxide synthase to plasmalemmal caveolae. *J. Biol. Chem.*, **271**, 6518–6522.

Smart, E.J., Ying, Y., Mineo, C. and Anderson, R.G.W. (1995). A detergent free method for purifying caveolae membrane from tissue cultured cells. *Proc. Natl. Acad. Sci. USA*, **92**, 10104–10108.

Song, K.S., Li, S., Okamoto, T., Quilliam, L.A., Sargiacomo, M. and Lisanti, M.P. (1996). Co-purification and direct interaction of Ras with caveolin, an integral membrane protein of caveolae microdomains: Detergent-free purification of caveolae membranes. *J. Biol. Chem.*, **271**, 9690–9697.

Song, K.S., Scherer, P.E., Tang, Z-L., Okamoto, T., Li, S., Chafel, M., Chu, C., Kohtz, D.S. and Lisanti, M.P. (1996). Expression of caveolin-3 in skeletal, cardiac and smooth muscle cells: Caveolin-3 is a component of the sarcolemma and co-fractionates with dystrophin and dystrophin-associated glycoproteins. *J. Biol. Chem.*, **271**, 15160–15165.

Tang, Z.-L., Scherer, P.E., Okamoto, T., Song, K., Chu, C., Kohtz, D.S., Nishimoto, I., Lodish, H.F. and Lisanti, M.P. (1996). Molecular cloning of caveolin-3, a novel member of the caveolin gene family expressed predominantly in muscle. *J. Biol. Chem.*, **271**, 2255–2261.

MOLECULAR DETERMINANTS OF AGONIST-INDEPENDENT ACTIVITY IN ADRENERGIC RECEPTORS

P. SAMAMA and R. J. LEFKOWITZ

Howard Hughes Medical Institute, Duke University Medical Centre, Durham, NC 27710, USA

Stimulation of adrenergic receptors by agonists results in the activation of G proteins and downstream effectors. *In vivo*, circulating catecholamines generate a signaling tone. However, expression of receptors in heterologous systems has recently led to the appreciation of ligand-independent signaling as an intrinsic property of many receptor subtypes. The biological importance of this activity suggests a role for inverse agonists in therapy and research. Constitutively activating mutations of receptors have helped define basal activity as an aspect of normal receptor function, and emphasize the existence of an activated conformation of the receptor. Such mutations are presumed to mimic the molecular rearrangements underlying receptor activation. Thus, the molecular basis of basal activity of receptors can shed light on their activation by agonists.

KEY WORDS: basal activity, constitutively active, inverse agonist

INTRODUCTION

Epinephrine and norepinephrine exert their actions by binding to cell-surface receptors on their target cells. The adrenergic receptors, to which they bind, are amongst the best-studied receptors which couple to G proteins. Signaling is initiated by the binding of epinephrine or norepinephrine to a site in contact with the extracellular space. This signal is then transmitted towards the intracellular face of the membrane, to cytoplasmic regions of the receptor molecule which in turn activate the cognate intracellular G protein. Thus, the receptor molecule transduces the hormonal signal by coupling two separate binding sites situated on different faces of the cell membrane.

Two models have been developed separately to account for receptor function (Figure 1). The allosteric model of Monod, Wyman and Changeux (1965), originally applied to hemoglobin and later to cell surface receptors (Karlin, 1967; Thron, 1973; Leff, 1995), focuses on the receptor protein. It postulates the existence of distinct conformational states in equilibrium, endowed with varying binding affinities for ligands. The binding of ligands, by selecting for a subset of conformations, alters the abundance of the various conformations, resulting in the biological effect (Burgen, 1981). The model also predicts the spontaneous occurrence of active forms, leading to a basal "tone", or ligand-independent receptor activity. One attraction of the allosteric model is its parsimony. Indeed, at

---

Correspondence to: P. Samama, Howard Hughes Medical Institute, Duke University Medical Centre, Durham, NC 27710, USA. Tel: 919-684-2974; Fax: 919-684-8875

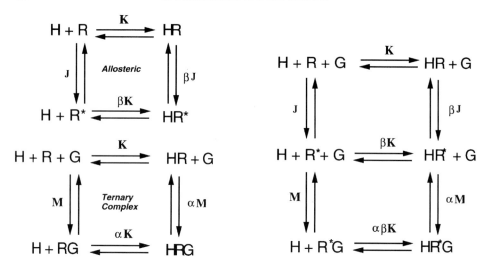

FIGURE 1 The Allosteric (top left), Ternary Complex (bottom left), and combined Allosteric-Ternary Complex (right) models of receptor action. H, ligand; R, R*, inactive and active conformations of the receptor. G, G protein. K, M, binding constants; β, α, dimensionless proportionality constants (efficacy). For agonists, β or α > 1 (see Samama et al., 1993).

its simplest, the model only requires two freely interconvertible states ("active" and "inactive") to account for pharmacological agonism. The "active" conformation is defined simultaneously by its biological effect, and by its greater avidity for agonists. Antagonists are predicted to fall into two classes: "neutral" blockers have equal avidity for both conformations, while "inverse agonists" bind preferentially to the inactive conformation. The nature of these conformations, however, is in general hypothetical.

In contrast, the Ternary Complex Model (De Léan, Stadel & Lefkowitz, 1980) incorporates biochemical knowledge of receptor-effector coupling mechanisms, in particular the existence of a diffusible G protein in the membrane plane. It emphasizes the reciprocal effects of agonist and G protein binding to the receptor, and was designed to explain the correlation of drug binding properties with drug activity. Indeed, this model accounts satisfactorily for the existence of biphasic, GTP-sensitive competition binding isotherms for agonists in a variety of broken cell preparations.

The allosteric model of receptor activation was widely applied to channel-type receptors, because single channel recording demonstrated the existence of discrete states of the molecule (open and closed), together with the existence of spontaneous channel opening in the absence of ligand (Hess, Lansman & Tsien, 1984). In contrast, for G protein-coupled receptors (with the notable exception of Rhodopsin), no such discrete behavior is apparent, and there is a continuous range of agonist efficacy. Thus, the Ternary Complex Model has been preferred whenever G protein-coupled receptors were studied. The two models, however, are by no means exclusive. Indeed, it is *implicit* in the TCM that the agonist-bound receptor somehow adopts a specific, active conformation (why else would it bind the G protein?). Moreover, the TCM allows for receptor pre-coupling, i.e. formation of RG complexes in the absence of ligand. It is the scope of this review to show how the study of ligand-independent activity, guided by these two models, sheds some light on receptor function.

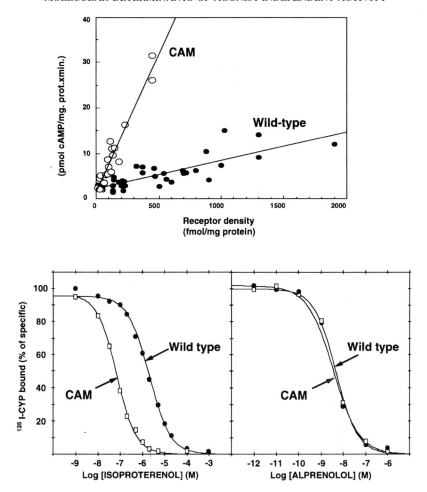

FIGURE 2 Properties of the constitutively active mutant β_2-AR: Top: basal activity of the wild-type and mutant receptors transfected in CHO cells. Bottom: competition binding of the antagonist alprenolol or the agonist isoproterenol against the radioligand ^{125}I-CYP (see Samama et al., 1993).

HOW GAIN-OF-FUNCTION MUTATIONS ILLUMINATE RECEPTOR ACTIVATION

In a study investigating the molecular determinants of the α_{1B}-AR coupling to G_q, Cotecchia and colleagues mutated the receptor in the distal region of the third intracellular domain. This mutant exhibited high agonist-independent signaling activity and high affinity for the agonist norepinephrine, but not for the antagonist prazosin (Cotecchia et al., 1990). A pattern of phenotypes was further established when the G_i-coupled α_2-AR and the G_s-coupled β_2-AR were also mutagenized, with similar results (Ren et al., 1993; Samama et al., 1993; Figure 2). Similar results were obtained for mutations in the same region in the m_1 muscarinic acetylcholine receptor (Högger et al., 1995), the LH receptor (Latronico et al., 1995) and the TSH receptor (Parma et al., 1993). Most intriguing was the fact that the mutations affected agonist-, but not antagonist binding. This property was addressed with the β_2-AR mutant, for which a variety of ligands with varying efficacies were available, and where the contribution of G proteins to binding properties could be easily determined. We found that, for a range of weak partial, partial and full agonists, the mutation increased the apparent binding affinity of drugs in a fashion correlated with their efficacy, and not dependent on G protein binding. As this could not be explained simply within the framework of the TCM, we sought to expand the model by introducing an *explicit* isomerization step for the receptor molecule. A hypothetical active conformation R^* would be responsible for G protein activation (Figure 1). It was further assumed that the mutation primarily favored the formation of the high-affinity, active R^* form at the expense of the low-affinity, inactive form. Computer simulations of dose-response and ligand binding curves demonstrated that the new composite "Allosteric-Ternary Complex model" described the properties of the CAM ARs satisfactorily (Samama *et al.*, 1993).

While the allosteric TCM was designed to account for gain-of-function mutations, not all such mutations published can be explained by an alteration of an $R \leftrightarrow R^*$ equilibrium, as the following examples show: (i) an enrichment in the R^* form is expected to produce an increase in apparent agonist affinity, a property not apparent for constitutively activated mutants of the LH receptor (Kosugi *et al.*, 1995). (ii) Similarly, a receptor with an abnormal level of R* should retain some responsiveness for the agonist (unless it exists only in the R^* form), contrary to what is observed for an MSHR mutant (Robbins *et al.*, 1993). (iii) More strikingly, when all possible substitutions for the residue 293 were examined at the α_{1B}-AR, a reasonable correlation was observed between the two readouts of the gain-of-function mutation: increase in apparent affinity for epinephrine, and increase in agonist-independent signalling (Kjelsberg *et al.*, 1992; Figure 3). Such a cor-

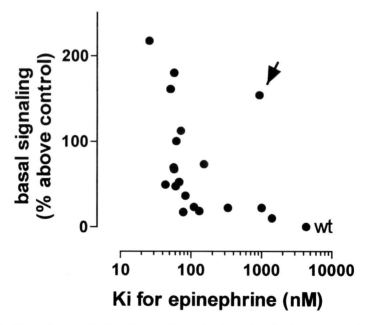

FIGURE 3 Changes in agonist binding affinity and basal signaling activity for nineteen substitutions at residue 293 of the α_{1B}-AR (see Kjelsberg *et al.*, 1992). Each point represents a mutant. The arrow indicates the $A_{293} \rightarrow C$ mutant.

relation is expected, as the $R \leftrightarrow R^*$ equilibrium would affect both agonist affinity and basal activity. However, at least one mutant, the Ala293 → Cys, exhibited strong basal activity, while showing a modest increase in affinity. Thus, this receptor mutant achieves high basal signaling in a way that must differ from the other mutants engineered at the same position. Hence, it should be kept in mind that the two-state model cannot account for the effect of all gain-of-function mutations. Rather, its scope is to provide a general hypothesis for the activation of a variety of wild-type G protein-coupled receptors by their respective ligands: namely, that receptor activation is governed by the equilibrium between an active (R^*) and an inactive (R) conformation of the receptor, and that agonists favor R^*, inverse agonists favor R, and neutral blockers have no preference.

GENERALITY OF BASAL ACTIVITY

Although basal activity is best revealed by mutated receptors, it is an intrinsic property of wild-type receptors as well. When overexpressed in cultured cells, receptors can exhibit a significant ligand-independent signaling activity. This is most strikingly illustrated by overexpression of the β_2-AR in the heart of transgenic mice, under the control of a cardiac-specific promoter. In the heart, the β_2-AR couples primarily to the activation of adenylyl cyclase. When the β_2-AR is overexpressed ~1000-fold (to 20–40 pmol/mg membrane protein), both basal and epinephrine-stimulated cyclase activities increase. Because of signal amplification, contractility, the endpoint of adrenergic stimulation in the heart, is maximally elevated (Milano *et al.*, 1994; Figure 4). This high basal tone can be decreased using inverse agonists (Bond *et al.*, 1995). It was crucial to demonstrate that the high basal activity did not result from the combined effects of high receptor expression levels (and therefore high sensitivity) with circulating catecholamines. Indeed, basal tone was not affected by reserpine treatment of the animals, which depletes tissue stores of catecholamines, and was lowered to a *varying* extent by different blockers (Bond *et al.*, 1995). This latter observation rules out that baseline tension is caused by marginal occupation of the β_2-AR by catecholamines. Thus, there is an important reciprocity between receptor basal activity and inverse agonism: while high basal activity is necessary for the observation of inverse agonism, inverse agonists of varying efficacies validate the observed signaling tone as truly ligand-independent.

Whether basal activity of wild-type receptors has functional relevance *in vivo* is as yet a matter of speculation. *In vitro*, different receptor subtypes can exhibit varying, characteristic levels of agonist-independent signaling activity (Tiberi & Caron, 1994). Thus, this activity might be part of the functional identity of receptor subtypes, together with specificity towards G proteins and patterns of spatial and temporal expression. *In vivo*, however, agonist-independent receptor activity is likely to be overshadowed by agonist-induced activity, such as that due to the low endogenous levels (nM range) of circulating catecholamines. Again, it may be possible to address this issue using inverse agonists. In contrast, other G-protein coupled receptors activated by mutations result in clearly pathological manifestations, such as tumors or hormonal dysfunction (Lefkowitz, 1993). In such pathologies, a clinically effective inverse agonist would be desirable. The fact that gain-of-function mutations of receptors have a clinical phenotype indicates that counter-regulatory mechanisms were not effective. It is intriguing that continuous,

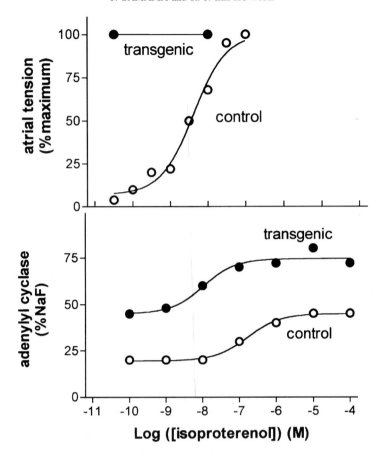

FIGURE 4 Effect of overexpression of the β_2-AR in the hearts of transgenic mice. Lower panel: adenylyl cyclase activity in particulate fractions of whole hearts. Top panel: atrial contractility *ex vivo* (see Milano *et al.*, 1994, and Bond *et al.*, 1995).

low-level activation of second-messenger pathways does not result in a potent attenuation of the signaling by mutant receptors. Indeed, we have demonstrated in transfected cells that the CAM β_2-AR engages down-regulation and desensitization mechanisms, in a fashion similar to the agonist-stimulated wild-type receptor (Pei *et al.*, 1994). It thus seems likely that a balance is achieved between tonic activity and tonic attenuation of the activity of these mutated receptors.

STRUCTURAL DETERMINANTS OF RECEPTOR ACTIVATION

In the current allosteric-ternary conception of receptor action, agonist binding changes the proportion, not necessarily the nature, of receptor conformations. Therefore, the active conformation responsible for agonist-independent activity need not be different from that responsible for agonist-induced activity. As a starting hypothesis, we can assume that the mechanism by which mutations enhance the constitutive activity of receptors provides insight into the molecular rearrangements which underlie agonist action.

The tertiary structure of adrenergic receptors is not known directly, but a large body of data suggests the existence of a seven-spanning transmembrane bundle of α-helices. The borders, positioning and orientation of the transmembrane segments are inferred from patterns of sequence conservation among related receptors, and from mutagenesis and cross-linking experiments. This information can then be incorporated into 3-D models of the transmembrane bundles (Baldwin, 1993; Ballesteros & Weinstein, 1995). Molecular dynamics simulations can be performed by computers, to determine how the binding of a ligand in the binding site affects molecular movements in the receptor molecule. Using such an approach, various groups have proposed mechanisms by which ligand binding promotes movements of transmembrane helices, transmitted in turn to intracellular hydrophilic regions which activate the G protein. For instance, Kontoyianni & Lybrand (1993) proposed that binding of epinephrine to the β_2-AR promotes a rotation of TM5, which is then transmitted to the third intracellular domain. Zhang & Weinstein (1993) have documented in detail how the binding of serotonin to the 5-HT$_2$ receptor is predicted to cause TM7 to "kink out" of the transmembrane bundle. The latter simulations emphasize the role of Proline residues as "hinges" within the transmembrane segments. In the light of the molecular dynamics experiments, two leading, non-exclusive hypotheses can be formulated for the activation of receptors: (i) perturbation of packing between the side chains of neighboring α-helices; this could be achieved, for instance, by rotational movements in the transmembrane bundle, resulting in an opening of the tertiary structure of the receptor, (ii) kinking of transmembrane helices around Proline residues, causing the lower part of the helices to protrude, perhaps acting as a trigger for G protein activation. These hypotheses can be investigated experimentally by mutagenesis of receptors. One key approach will be to search for compensatory mutations for the gain-of-function mutations in adrenergic receptors. The pattern of such compensatory mutants will help probe the intramolecular interactions which normally constrain the inoccupied receptor to an inactive state. Thus, beyond emphasizing the meaning of agonist-independent activity, these mutants will help to investigate the processes of agonist activation of G protein-coupled receptors.

ABBREVIATIONS

AR, Adrenergic Receptor; LH, Luteinizing Hormone, MSH, Melanocyte-Stimulating Hormone, FSH, Follicle-Stimulating Hormone, TCM, Ternary Complex Model; CAM, Constitutively Active Mutant; TM, Transmembrane segment.

References

Baldwin, J.M. (1993). The probable arrangement of the helices in G protein-coupled-receptors. *EMBO J.*, **12**, 1693–1703.

Ballesteros, J.A. and Weinstein, H. (1995). Integrated methods for the construction of three-dimensional models and computational probing of structure-function relations in G protein-coupled receptors. In *Methods in Neurosciences, Vol. 25*, edited by S. Sealfon, pp. 366–428. Academic Press. Inc.

Bond, R.A., Leff, P., Johnson, T.D., Milano, C.A., Rockman, H.A., McMinn, T.R., *et al.* (1995). Physiological effects of inverse agonists in transgenic mice with myocardial overexpression of the β_2-adrenoreceptor. *Nature*, **374**, 272–276.

Burgen, A.S.V. (1981). Conformational changes and drug action. *Fed. Proc.*, **40**, 2723–2728.
Cotecchia, S., Exum, S., Caron, M.G. and Lefkowitz, R.J. (1990). Regions of the α_1-adrenergic receptor involved in coupling to phosphatidylinositol hydrolysis and enhanced sensitivity to biological function. *Proc. Natl. Acad. Sci. USA*, **87**, 2896–2900.
De Léan, A., Stadel, J.M. and Lefkowitz, R.J. (1980). A ternary complex model explains the agonist-specific binding properties of the adenylate cyclase-coupled β-adrenergic receptor. *J. Biol. Chem.*, **255**, 7108–7117.
Hess, P., Lansman, J.B. and Tsien, R.W. (1984). Different modes of Ca channel gating behaviour favored by dihydropyridine Ca agonists and antagonists. *Nature*, **311**, 538–544.
Högger, P., Shockley, M.S., Lameh, J. and Sadée, W. (1995). Activating and inactivating mutations in the N- and C-terminal I3 loop junctions of muscarinic acetylcholine Hm1 receptors. *J. Biol. Chem.*, **270**, 74705–74710.
Karlin, A. (1967). On the application of "a plausible model" of allosteric proteins to the receptor for acetylcholine. *J. Theoret. Biol.*, **16**, 306–320.
Kjelsberg, M.A., Cotecchia, S., Ostrowski, J., Caron, M.G. and Lefkowitz, R.J. (1992). Constitutive activation of the α_{1B}-adrenergic receptor by all amino-acid substitution at a single site. *J. Biol. Chem.*, **267**, 1430–1433.
Kontoyianni, M. and Lybrand, T.P. (1993). Computer modeling studies of G protein-coupled receptors. *Med. Chem. Res.*, **3**, 407–418.
Kosugi, S., Van Dop., C., Geffner, M.E., Rabl, W., Carel, J.-C., Chaussain, J.-L., et al. (1995). Characterization of heterogenous mutations causing constitutive activation of the luteinizing hormone receptor in familial male precocious puberty. *Hum. Mol. Gen.*, **4**, 183–188.
Latronico, A., Anasti, J., Arnhold, I.J.P., Mendonça, B.B., Domenice, S., Albano, M.C., Zachman, K., et al. (1995). A novel mutation of the luteinizing hormone receptor gene causing male gonadotropin-independent precocious puberty. *J. Clin. Endocrinol. Metabol.*, **80**, 2490–2494.
Leff, P. (1995). The two-state model of receptor activation. *Trends in Pharmacol. Sci.*, **16**, 89–97.
Lefkowitz, R.J. (1993). Turned on to ill effect. *Nature*, **365**, 603–604.
Milano, C.A., Allen, L.F., Rockman, H.A., Dolber, P.C., McMinn, T.R., Chien, K.R., et al. (1994). Enhanced myocardial function in transgenic mice overexpressing the β_2-adrenergic receptor. *Science*, **264**, 582–586.
Monod, J., Wyman, J. and Changeux, J.P. (1965). On the nature of allosteric transitions: a plausible model. *J. Mol. Biol.*, **12**, 88–118.
Parma, J., Duprez, L., Van Sande, J., Cochaux, P., Gervy, C., Mockel, J., et al. (1993). Somatic mutations in the thyrotropin receptor gene cause hyperfunctioning thyroid adenomas. *Nature*, **365**, 649–651.
Pei, G., Samama, P., Lohse, M., Wang, M., Codina, J. and Lefkowitz, R.J. (1994). A constitutively active mutant β_2-adrenergic receptor is constitutively desensitized and phosphorylated. *Proc. Natl. Acad. Sci. USA*, **91**, 2699–2702.
Ren, Q., Kurose, H., Lefkowitz, R.J. and Cotecchia, S. (1993). Constitutively active mutants of the α_2-adrenergic receptor. *J. Biol. Chem.*, **268**, 16483–16487.
Robbins, L.S., Nadeau, J.H., Johnson, K.R., Kelly, M.A., Roselli-Rehfuss, L., Baack, E., et al. (1993). Pigmentation phenotypes of variant extension locus alleles result from point mutations that alter MSH receptor function. *Cell*, **72**, 827–834.
Samama, P., Cotecchia, S., Costa, T. and Lefkowitz, R.J. (1993). A mutation-induced activated state of the β_2-adrenergic receptor. *J. Biol. Chem.*, **268**, 4625–4636.
Tiberi, M. and Caron, M.G. (1994). High agonist-independent activity is a distinguishing feature of the dopamine D_{1B} receptor subtype. *J. Biol. Chem.*, **269**, 27925–27931.
Thron, C.D. (1973). On the analysis of pharmacological experiments in term of an allosteric model. *Mol. Pharmacol.*, **9**, 1–9.
Zhang, D. and Weinstein, H. (1993). Signal transduction by a 5-HT$_2$ receptor: a mechanistic hypothesis from molecular dynamics simulations of the three-dimensional model of the receptor complexed to ligands. *J. Med. Chem.*, **36**, 934–938.

α_2 ADRENERGIC RECEPTOR ACTIVATION AND INTRACELLULAR Ca^{2+}

KARL E. O. ÅKERMAN, CARINA I. HOLMBERG, HUI-FANG GEE, ANNIKA RENVAKTAR, SANNA SOINI, JYRKI P. KUKKONEN, JOHNNY NÄSMAN, KRISTIAN ENKVIST, MICHAEL J. COURTNEY and CHRISTIAN JANSSON

Department of Biochemistry and Pharmacy, Åbo Akademi University, Biocity, P.O. Box 66, FIN-20521 Turku, Finland

The α_2 adrenergic receptors couple to multiple signal transduction pathways and may induce either stimulatory or inhibitory responses. An increase in cytosolic free Ca^{2+} may explain many of the stimulatory effects. When expressed in PC-12 cells these receptors stimulate Ca^{2+} currents by a mechanism which is insensitive to pertussis toxin while inhibition of Ca^{2+} currents mediated by these receptors is sensitive to the toxin. In HEL erytroleucemia cells and in astrocytes endogenous α_2 receptor activation increase cytosolic free Ca^{2+} by a pertussis toxin sensitive mechanism which appears to be mediated by phospholipase C (PLC). Recombinant α_2 receptors in CHO cells or Sf9 insect cells mediate similar responses. It is suggested that G_s and G_i type G-proteins respectively are responsible for stimulation and inhibition of Ca^{2+} currents while receptor mediated Ca^{2+} mobilisation is due to activation of PLC through $\beta\gamma$ subunits of G-proteins (mainly G_i).

KEY WORDS: α_2-adrenoceptors, calcium, IP_3, recombinant receptors, endogenous receptors

INTRODUCTION

Depending on the tissue where they are expressed α_2-adrenergic receptors may induce either inhibitory or stimulatory responses apparently through different mechanisms. Central inhibitory responses include behavioural depression, sedation and analgesia. These effects are thought to be mediated through an activation of inhibitory pre- or postsynaptic receptors on neurons. Peripheral inhibitory responses to α_2-adrenergic receptor activation include inhibition of insulin release, water absorption, reduction in gastrointestinal motility and secretion as well as inhibition of lipolysis. The primary stimulatory responses known to be mediated by α_2-adrenergic receptors include platelet aggregation and contraction of smooth muscle as well as stimulation of secretion in some cells.

CELLULAR MECHANISMS OF THE INHIBITORY RESPONSES

Inhibition of adenylyl cyclase mediated through G_i-type G-proteins is typically seen upon α_2-adrenergic receptor activation. Inhibition of cAMP production is also usually seen

---

Correspondence to: Karl E.O. Åkerman, Dept. of Biochemistry and Pharmacy, Åbo Akademi University, Biocity, P.O. Box 66, FIN-20521 Turku, Finland

when recombinant α_2-adrenergic receptor subtypes α_{2A}, α_{2B} or α_{2C} are expressed in target cells (Limbird, 1988) and could be the basis of some of the inhibitory effects mediated through these receptors. Voltage gated Ca^{2+} channels are inhibited by α_2-receptors in neuronal cells (Song et al., 1991; Gollasch et al., 1991; Bley and Tsien, 1990). These responses are transduced through the pertussis toxin-sensitive G_i/G_o type of G-proteins.

CELLULAR MECHANISMS OF STIMULATORY RESPONSES

Stimulation of cAMP production by α_2-adrenergic receptors

In different target cells (Fraser et al., 1989; Jones et al., 1991; Duzic and Lanier, 1992; Jansson et al., 1994; Jansson et al., 1995) α_2-adrenergic receptors increase cAMP production. In other cell lines no stimulation of cAMP production has been seen (Coteccia et al., 1990; Eason et al., 1992; Duzic and Lanier, 1992; Jansson et al., 1994) partially dependent on the receptor subtype expressed. The α_{2B} subtype in particular seems to couple primarily to stimulation in some cells including PC-12 cells (Duzic and Lanier, 1992) S115 cells (Jansson et al., 1994), JEG-3 cells (Pepperl and Regan, 1993) and Sf9 insect cells (Jansson et al., 1995). The stimulation of cAMP production by α_2 receptors is not mediated by pertussis toxin sensitive G-proteins and different types of second messenger-induced changes in cAMP levels have been excluded (Jones et al., 1991; Jansson et al., 1994). In fact a direct interaction between α_2 receptors and G_s has been demonstrated (Eason et al., 1992) suggesting that these receptors may couple through at least two G-proteins.

Coupling of α_2-adrenergic receptors to intracellular Ca^{2+} changes

Regulation of Ca^{2+} elevation by α_2-receptors is one of the more recently discovered functions. Ca^{2+} elevation elicited by α_2-receptors has also been associated with contraction of some smooth muscle, based on inhibition of contraction in the absence of external Ca^{2+} (Nielsen et al., 1992), and blocking of contractile responses by blockers of voltage gated Ca^{2+} channels. These results have indicated that α_2 receptors activate voltage gated Ca^{2+} channels of the L-type while inhibition (see above) mainly occurs through N-type ω-contoxin sensitive Ca^{2+} channels. Recently, the coupling of the α_2-adrenergic receptors to an increase in Ca^{2+} in smooth muscle has been confirmed by direct Ca^{2+} measurements (Aburto et al., 1993; Erdbrügger et al., 1993).

Ca^{2+} elevations have also been measured in human erythroleukemia (HEL) cells (Michel et al., 1989) and in human platelets (Kagaya et al., 1992) which should express the α_{2A} receptor subtype. In these cells Ca^{2+} mobilisation occurred partially from internal stores and was sensitive to pertussis toxin suggesting that this response like the inhibition of cAMP production is mediated by the G_i/G_o type G-proteins. Intracellular Ca^{2+} mobilisation is usually connected to the activation of phosphoinositolphospholipid (PI) specific phospholipase C (PLC) leading to production of inositol 1,4,5-triphosphate (IP_3) which binds to a receptor in membrane of endoplasmatic reticulum leading to an opening of a Ca^{2+} channel and a subsequent rise in cytosolic Ca^{2+} (Berridge, 1987).

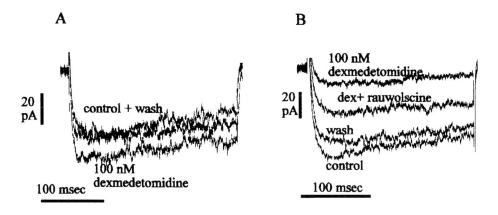

FIGURE 1 Effect of the α_2 receptor agonist on Ca^{2+} currents in NGF treated PC-12 pheochromocytoma cells. Cells were treated with 50 ng/ml NGF for 10 days. Whole cell currents were recorded from a holding potential of –80 mV followed by a voltage step to –10 mV. **A**. One cell showing a stimulation of Ca^{2+} currents. **B**. A cell showing an inhibitory response. The concentration of rauwolscine was 1 µM. **Below** results collected from several similar experiments. Pertussis toxin treatment was performed using 100 mg/ml for 2 days. The electrode solution contained (in mM): 120 CsCl, 3 $MgCl_2$ 5 Mg-ATP, 0.1 GTP 10 EGTA, 5 hepes pH 7.4 and the external solution contained (in mM): 125 NaCl, 10.8 $BaCl_2$, 1 $MgCl_2$ 5.4 CsCl 10 glucose and 10 Hepes pH 7.4.

Coupling of α_{2D} and α_{2B} receptor subtypes to Ca^{2+} channels in PC-12 pheochromocytoma cells. Differentiated PC-12 pheochromocytoma cells are often used as models for neurons. These cells differentiate towards a more neuronal phenotype when treated with NGF. Expressing recombinant α_2-receptors in these cells does not change this property. The cells grow long neuritelike processes. Ca^{2+} currents were recorded from these cells using the whole cell patch clamp method. Figure 1 shows recordings from two cells expressing the α_{2D} receptor. Surprisingly stimulation of α_2-adrenergic receptors with dexmedetomidine induces either a stimulatory or an inhibitory response. In Figure 1A an increase of the magnitude of Ca^{2+} current is seen which is reversed by washing out the agonist. In Figure 1B a substantial inhibition of the Ca^{2+} current is seen which is partially reversed by the α_2 receptor antagonist rauwolscine and totally reversed by washing out the drugs. A stimulatory response was seen in 5–20% of the cells while about 50% showed an inhibi-

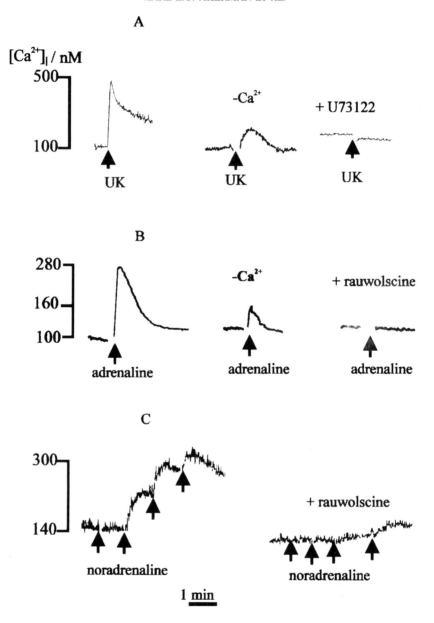

FIGURE 2 Effect of α_2 receptor agonists and antagonists on intracellular free Ca^{2+} in HEL cells CHO cells and Sf9 cells. **A**. Fura-2 fluorescence from HEL cell suspensions was recorded using a Hitachi F4000 fluorescence spectrophotometer in a stirred microcuvette placed in a thermostated holder in a medium containing (in mM): 137 NaCl, 5 KCl, 1 $CaCl_2$ 1.2 $MgCl_2$, 0.44 KH_2PO_4, 4.2 $NaHCO_3$, 10 glucose 20 TES, pH 7.4. As indicated 1 µM UK14,304 was added in the absence or presence of 5 mM EGTA or 10 µM U73122. **B**. Recordings from CHO cells (on coverslips placed in a cuvette) expressing the α_{2B} receptor. Adrenaline at 1 µM was added in the presence or absence of EGTA as above or in the presence of 1 µM rauwolscine as indicated. **C**. Sf9 cells infected with recombinant baculovirus containing the α_{2B} receptor. Additions were made of noradrenaline in increasing concentrations (10 nM, 110 nM, 1.1 µM, 11 µM and 110 µM) as indicated by the arrows and on the right a similar experiment is shown performed in the presence of 28 nM rauwolscine.

tory response. Pretreatment of the cells with pertussis toxin significantly reduced the proportion of cells showing an inhibitory response (10–20%) and increased the stimulatory proportion (30–80%). The α_{2B} subtype expressing cells showed far more stimulatory responses than those expressing α_{2A}. Thus recombinant α_2 receptors are able to mediate both an inhibition and stimulation of Ca^{2+} currents in PC-12 cells.

Ca^{2+} mobilisation by endogenously expressed α_2 receptors in HEL erythroleucemia cells is blocked by the PLC inhibitor U73122. It has been shown previously (Michel et al., 1989) that α_2 receptor activation in HEL erythroleucemia cells induces Ca^{2+} mobilisation as determined with the fluorescent probe fura-2 (Grynkiewicz et al., 1985). Typical fura-2 recordings from these cells are shown in Figure 2A. Addition of the α_2 receptor agonist UK14,304 to cells that had been prelabelled with fura-2 leads to a significant increase in the fluorescence of the entrapped dye indicating a rise in cytosolic Ca^{2+}. As usually seen in case of PLC mediated responses (Berridge, 1987) the signals consists of a fast transient phase thought to be intracellular release and a slow sustained phase due to influx from the external medium. If Ca^{2+} is removed from the external solution by addition of EGTA prior to the addition of agonist a smaller and more transient response is seen. In order to test whether the response is connected to the activation of PLC a blocker of this enzyme U73122, was used. This compound totally inhibited the response to UK.

Ca^{2+} mobilisation through recombinant α_2 receptor activation. In Figure 2B fura-2 recordings from CHO cells, on coverslips, expressing recombinant α_{2B} receptors is shown. Addition of adrenaline causes an increase in fura-2 fluorescence similar to the response in HEL cells. In the absence of extracellular Ca^{2+} the signal is smaller and more transient. The α_2 receptor antagonist rauwolscine completely blocks the response suggesting that it is due to α_{2B} receptor activation. Qualitatively similar responses were seen with cells expressing the α_{2C} receptor. The magnitude of these responses was, however, considerably smaller (data not shown).

The baculovirus expression system has been used for high yield production of various recombinant proteins. We have previously shown, using this system, that α_2 receptors are functionally coupled to the cAMP second messenger systems in Sf9 insect cells (Oker-Blom et al., 1993; Jansson et al., 1995). In Figure 2C it is shown that in Sf9 cells infected with recombinant viruses expressing the α_{2B} receptor an increase in fura-2 fluorescence is seen upon addition of noradrenaline. The responses in these cells are different from those usually seen in mammalian cells in that they are slow and much more sustained. This makes it possible to perform dose-response measurements by sequential additions of increasing concentrations of agonists or antagonists. The difference in response kinetics probably reflects differences in the Ca^{2+} regulation in the cells rather than differences in receptor function. Similar slow and sustained responses are seen when muscarinic m1, m3 or m5 receptors are expressed using this system (data not shown). Addition of rauwolscine prior to noradrenaline considerably attenuates the Ca^{2+} signals (Figure 2C). When expressed with this system activation of α_{2A} and α_{2C} receptor subtypes also increased the fura-2 fluorescence but considerably less than by the α_{2B} receptor.

Induction of Ca^{2+} mobilisation by α_2 receptor activation in primary cultures of astrocytes by cAMP analogues. An increase in intracellular free Ca^{2+} upon α_2-receptor stimulation

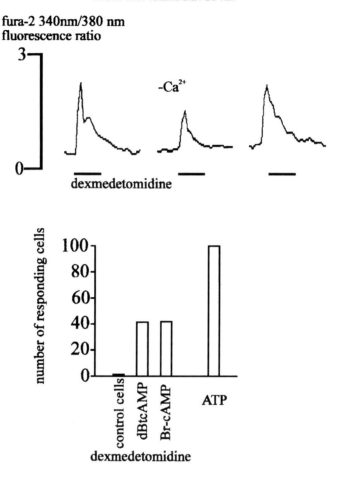

FIGURE 3 Effect of α_2 receptor agonists dexmedetomidine on intracellular free Ca^{2+} in cultured astrocytes. Cells were grown for three days in the presence of 1 mM dibutyryl cAMP on round ($\varnothing$ = 22 mm) coverslips. The coverslip was placed in a perfusion chamber in a Nikon Diaphot microscope. The fura-2 loaded cells were excited through alternating 340 nm/380 nm filters and the fluorescence at 505 nm was monitored with a SIT camera. Images were collected every 3 sec and were digitised using a SPEX DM3000 image analyser as described in detail earlier (Lindqvist et al., 1995). The images were subsequently analysed by a software package developed by Dr M.J. Courtney. The ratio of 340 nm/380 nm fluorescence from a single cell is shown in response to 10 µM dexmedetomidine in the absence or presence of EGTA.

has previously been shown, in single cell measurements on primary cultures of astrocytes (Salm and McCarthy, 1990). Using fura-2 and image analysis such a response was seen only in very few cells (about 1% depending on the cell batch). Treatment of these cells with permeable cAMP analogues leads to a phenotype resembling reactive astrocytes. With this treatment the number of cells responding to the α_2 receptor agonist dexmedetomidine increases significantly (Figure 3). As was the case with HEL and CHO cells a response although smaller in magnitude is seen after Ca^{2+} removal from the external medium. RNAse protection analysis of receptor subunit mRNA expression indicated that the α_{2A} subtype is induced in these cells by the cAMP analogues (data not shown).

IP$_3$ production by α_2 receptor activation. It has previously been shown that recombinant α_{2A}- and α_{2C}-adrenoceptors can stimulate inositol phospholipid hydrolysis in some expression systems (Cotecchia et al., 1990). Stimulation of α_2 receptors (for 10–20 sec) causes an increase in the production of IP$_3$ (as measured using a kit from Amersham) in HEL (1.5 ± 0.11 fold) cells, astrocytes (1.9 ± 0.36 fold) and Sf9 (1.8 ± 0.09 fold) cells in conditions as in Figures 2 and 3.

CONCLUSIONS

Although Ca^{2+} has not previously been considered to be a primary signal in α_2-adrenergic receptor mediated responses it seems to have several connections to pathways coupling to Ca^{2+} as a messenger including actions on voltage gated Ca^{2+} channels and the PI system.

Coupling to Ca^{2+} channels. Inhibitory responses by α_2-receptors especially presynaptic ones are most likely to be due to blocking of voltage gated Ca^{2+} channels (Bley and Tsien, 1990; Song et al., 1991; Gollasch et al., 1991) mainly of the N-type (Bley and Tsien, 1990). Using recombinant α_2 receptors expressed in PC-12 cells we show here that the inhibitory responses are similar with α_{2D} and α_{2B} subtypes of receptors and are reduced by pertussis toxin treatment suggesting a coupling through G$_i$/G$_o$ type of G-proteins. The stimulation of Ca^{2+} currents was much more prominent in cells treated with pertussis toxin and was also more frequently seen with the α_{2B} subtype. The mechanism of the stimulatory response is unknown at present but could well be related to the stimulatory effect by α_2 receptors on L-type Ca^{2+} channels in smoth muscle (Nielsen et al., 1992). A direct stimulatory effect of α_s on L-type Ca^{2+} channels has also been shown (Yatani et al., 1988). Duzic and Lanier (1992) have previously shown a stimulation of cAMP production in these cells upon α_2 receptor activation. Taken together with the finding by Eason et al. (1992) that the increase in cAMP producution may be due to an interaction of G$_s$ with the receptor and several studies demonstrating a pertussis toxin insensitive increase in cAMP (Fraser et al., 1989; Jones et al., 1991; Duzic and Lanier, 1992; Jansson et al., 1994; Jansson et al., 1995; Pepperl and Regan, 1993) it appears possible that the inhibition of Ca^{2+} currents may be due to α_i and the stimulatory response may be through α_s subunits of the G-proteins (Figure 4) interacting with the channels.

Coupling of α_2 receptors to intracellular Ca^{2+} mobilisation. Results demonstrating α_2 receptor mediated Ca^{2+} mobilisation in HEL erythroleucemia cells (Michel et al., 1989) and human platelets (Kagaya et al., 1992) as well as results showing an increase in IP$_3$ production by stimulating recombinant receptors (Coteccia et al., 1990) may suggest that α_2 receptors are able to couple to the PI system. An activation of PLC is suggested by blocking the α_2 receptor mediated Ca^{2+} mobilisation by U73122, a blocker of PI specific PLC (Bleasdale et al., 1990). The production of IP$_3$ also indicate a stimulation of PLC. PLCβ is activated by α subunits of G$_{q/11}$ type G-protein but not by α_i, α_o or α_s (Hepler et al., 1993). Therefore a direct coupling of these subunits to phospholipase C is unlikely. As the Ca^{2+} mobilisations have been shown to be sensitive to pertussis toxin the most likely explanation is the activation of PLCβ by $\beta\gamma$ subunits (from G$_i$) (Camps et al., 1992). This hypothesis is schematically shown in Figure 4.

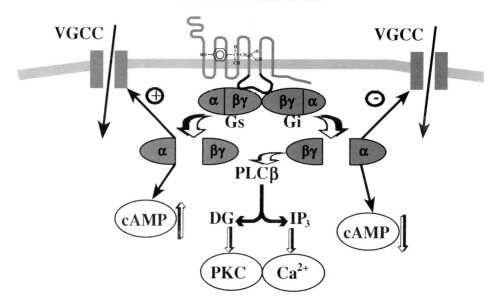

FIGURE 4 Working hypothesis to explain signal pathways leading to multiple responses through α_2 receptors. The α_2 interacts alternatively with G_s or G_i. It is proposed that α_i reduces the activity of N-type Ca^{2+} channels (VGCC) and adenylate cyclase and the $\beta\gamma$ subunits activate phospholipase Cβ producing diacylglycerol which activates protein kinase C (PKC) and inositol 1,4,5-triphosphate (IP$_3$) which mobilises Ca^{2+}. The α_s subunit again increases the activity of L-type Ca^{2+} channels (VGCC) and activates adenylyl cyclase.

ACKNOWLEDGEMENTS

The support by Sigrid Juse'lius Foundation, The Wellcome Trust and TEKES is greatly appreciated. PC-12 cells were generously provided by Dr S.M. Lanier and CHO cells by A. Marjamäki and M. Scheinin.

References

Aburto, T.K., Lajoie, C. and Morgan, K.G. (1993). Mechanism of signal transduction during α_2-adrenergic receptor-mediated contraction of vascular smooth muscle. *Circ. Res.*, **72**, 778–785.
Berridge, M.J. (1987). Inositol triphosphate and diacylglycerol: two interacting second messengers. *Annu. Rev. Biochem.*, **56**, 159–193.
Bley, K.R. and Tsien, R.W. (1990). Inhibition of Ca^{2+} and K^+ channels in sympathetic neurones by neuropeptides and other ganglionic transmitters. *Neuron*, **2**, 379–391.
Camps, M., Carozzi, A., Schnabel, P., Scheer, A., Parker, P. and Gierschik, P. (1992). Isoenzyme-selective stimulation of phospholipase C-β2 by G-protein $\beta\gamma$ subunits. *Nature*, **360**, 684–688.
Cotecchia, S., Kobilka, B.K., Daniel, K.W., Nolan, R.D., Lapetina, E.Y., Caron, M.G., Lefkowitz, R.J. and Regan, J.W. (1990). Multiple second messenger pathways of α-adrenergic receptor subtypes expressed in eukaryotic cells. *J. Biol. Chem.*, **265(1)**, 63–69.
Duzic, E. and Lanier, S.M. (1992). Factors determining the specificity of signal transduction by guanine nucleotide-binding protein-coupled receptors. III. Coupling of α_2-adrenergic receptor subtypes in a cell type-specific manner. *J. Biol. Chem.*, **267**, 24045–24052.
Eason, M.G., Kurose, H., Holt, B.D., Raymond, J.R. and Liggett, S.B. (1992). Simultaneous coupling of α_2-adrenergic receptors to two G-proteins with opposing effect. *J. Biol. Chem.*, **267**, 15795–15801.
Erdbrügger, W., Visher, P., Bauch, H.-J. and Michel, M.C. (1993). Norepinephrine and neuropeptide Y increase intracellular Ca^{2+} in cultured porcine aortic smooth muscle cells. *J. Cardiovasc. Pharmacol.*, **22**, 97–102.

Fraser, C.M., Arakawa, S., McComble, W.R. and Venter, J.C. (1989). Cloning, sequence analysis, and permanent expression of a human α_2-adrenergic receptor in Chinese hamster ovary cells. *J. Biol. Chem.*, **264**, 11754–11761.

Gollasch, M., Hescheler, J., Spicher, K., Klinz, F.-J., Schulz, G. and Rosenthal, W. (1991). Inhibition of Ca^{2+} channels via α_2-adrenergic and muscarinic receptors in pheochromocytoma (PC-12) cells. *Am. J. Physiol.*, **257**, C1282–1289.

Grynkiewicz, G., Poenie, M. and Tsien, R.Y. (1985). A new generation of Ca^{2+} indicators with greatly improved fluorescence properties. *J. Biol. Chem.*, **260**, 3440–3450.

Hepler, J.R., Kozasa, T., Smrka, V., Simon, M.I., Rhee, S.G., Sternweiss, P.C. and Gilman, A.G. (1993). Purification from Sf9 cells and characterization of recombinant $G_{q\alpha}$. Activation of purified phospholipase C isoenzymes by G_α subunits. *J. Biol. Chem.*, **266**, 14367–14375.

Jansson, C.C., Marjamäki, A., Luomala, K., Savola, J.-H., Scheinin, M. and Åkerman, K.E.O. (1994). Coupling of human α_2-adrenoceptor subtypes to regulation of cAMP production in transfected S115 cells. *Eur. J. Pharmacol.*, **266**, 165–174.

Jansson, C.C., Karp, M., Oker-Blom, C., Näsman, J., Savola, J.M. and Åkerman, K.E.O. (1995). Opposite coupling of two human α_2 adrenergic receptor subtypes α_{2A} and α_{2B}, expressed in Sf9 cells. *Eur. J. Pharmacol.*, **290**, 75–83.

Jones, B.S., Halenda, S.P. and Bylund, D.B. (1990). α_2-Adrenergic receptor stimulation of phosoholipase A_2 and of adenylyl cyclase in transfected Chinese hamster ovary cells is mediated by different mechanisms. *Mol. Pharmacol.*, **39**, 239–245.

Kagaya, A., Mikuni, M., Yamamoto, H., Muraoka, S., Yamaki, S. and Takahashi, ?. (1992). Heterologous supersensitization between serotonin$_2$ and α_2-adrenergic receptor-mediated intracellular calcium mobilization in human platelets. *J. Neural. Transm.*, **88**, 25–36.

Limbird, L.E. (1988). Receptors linked to inhibition of adenylate cyclase: additional signaling mechanisms. *FASEB J.*, **2**, 2686–2695.

Lindqvist, C., Holmberg, C., Oetken, C., Courtney, M.J., Ståhls, A. and Åkerman, K.E.O. (1995). Rapid Ca^{2+} mobilisation in single LGL upon interaction with K562 target cells. Role of CD18 and CD16 molecules. *Cell Immunol.*, **165**, 71–76.

Michel, M.C., Brass, L.F., Williams, A., Bokoch, G.M., LaMorte, V.J. and Motulsky, H.J. (1989). α_2-adrenergic receptor stimulation mobilizes intracellular Ca^{2+} in human erythroleukemia cells. *J. Biol. Chem.*, **9**, 4986–4991.

Nielsen, H., Mortensen, F.V., Pilegaard, H.K., Hasenkam, J.M. and Mulvany, M.J. (1992). Calcium utilization coupled to stimulation of postjunctional α_1 and α_2 adrenoceptors in isolated human resistence arteries. *J. Pharmacol. Exp. Ther.*, **260(2)**, 637–643.

Oker-Blom, C., Jansson, C., Karp, M., Lindqvist, C., Savola, J.-M., Vlak, J. and Åkerman, K.E.O. (1993). Functional analysis of the human α_{2C}-C4 adrenergic receptor in insect cells expressed by a luciferase-based baculovirus vector. *Biochim. Biophys. Acta*, **1176**, 269–275.

Salm, A.K. and McCarthy, K.D. (1990). Norepinephrine-evoked calcium transients in cultured cerebral type 1 astroglia. *Glia*, **3**, 529–538.

Song, S.-Y., Saito, K., Noguchi, K. and Konishi, S. (1991). Adrenergic and cholinergic inhibition of Ca^{2+} channels mediated by different GTP-binding proteins in rat sympathetic neurones. *Pflügers Arch.*, **418**, 592–60.

Yatani, A., Imoto, Y., Codina, J., Hamilton, S.L., Brown, A.M. and Birnbaumer, L. (1988). The stimulatory G protein of adenylyl cyclase, G_s, also stimulates dihydropyridine-sensitive Ca^{2+} channels. *J. Biol. Chem.*, **263**, 9887–9895.

THE INTERFACE BETWEEN α_2-ADRENOCEPTORS AND TYROSINE KINASE SIGNALLING PATHWAYS

GRAEME MILLIGAN, ANDREW R. BURT, MOIRA WILSON and NEIL G. ANDERSON[*]

Molecular Pharmacology Group, Division of Biochemistry and Molecular Biology, Institute of Biomedical and Life Sciences, University of Glasgow, Glasgow G12 8QQ, Scotland, UK and [*]Hannah Research Institute, Ayr KA6 5HL, Scotland, UK

Considerable interest has recently focused on observations that members of the family of G protein-coupled receptors can stimulate the activity of the extracellularly regulated kinases of the MAP kinase family of proteins. Early studies in this area centred on the activity of lysophosphatidic acid (LPA). However, because this receptor has not been identified at a molecular level and there is a very limited pharmacology at this site we selected the α_{2A}-adrenoceptor for such studies following stable transfection in Rat-1 fibroblasts. Clones expressing the α_{2A}-adrenoceptor were shown to mediate agonist-stimulation of high affinity GTPase activity, inhibition of forskolin-amplified adenylyl cyclase and, more surprisingly, stimulation of phospholipase D (PLD) activity.

Agonists at this receptor were also able to cause GTP loading of p21^{ras} and phosphorylation of both p42^{MAPK} and p44^{MAPK} isoforms. All these effects were prevented by prior treatment of the cells with pertussis toxin. Sustained elevation of intracellular cAMP levels by treatment with dibutyryl cAMP prevented α_{2A}-adrenoceptor-mediated activation of MAP kinase. Both LPA and epidermal growth factor also caused phosphorylation of p42^{MAPK} and p44^{MAPK} but only the effects of LPA was prevented by pertussis toxin treatment. Activation of the ribosomal S6 kinase p70^{S6K}, which is often considered to lie downstream of phosphoinositide-3 kinase, was also produced by agonist occupancy of the α_{2A}-adrenoceptor and this effect was prevented by both pertussis toxin treatment and by the macrolide rapamycin which functions via a rapamycin-associated protein (FRAP).

Analysis of clones expressing similar levels of the wild type and an Asp79Asn mutant of the α_{2A}-adrenoceptor indicated much reduced coupling efficiency of the mutant receptor to the cellular G protein population. This did not however attenuate the ability of agonists at the receptor to cause stimulation of p42^{MAPK} or p44^{MAPK} indicating that only a small fraction of the cellular G_i population is required to cause activation of these kinases.

KEY WORDS: adrenoceptor, MAP kinase, phosphoinositide-3 kinase, ribosomal S6 kinase, pertussis toxin, cyclic AMP.

INTRODUCTION

The α_2-adrenoceptors are the prototypic examples of seven transmembrane element, guanine nucleotide binding protein (G protein)-linked receptors which mediate their functions via interactions with members of the family of G_i-like heterotrimeric G proteins. In most physiological settings this is recorded as inhibition of adenylyl cyclase activity but within the CNS and in other excitable tissues this can also involve regulation of certain classes of ion channels such as voltage-operated Ca^{2+} channels.

Many non G protein-linked single trans-plasma membrane span receptors display intrinsic tyrosine kinase activity leading to the downstream activation of cascades of other

---

Correspondence to: Graeme Milligan, Division of Biochemistry and Molecular Biology, Institute of Biomedical and Life Sciences, University of Glasgow, Glasgow G12 8QQ, Scotland, UK. Fax: 44-141-330 4620

kinases which can involve phosphorylation of both tyrosine and serine/threonine residues in target proteins (Malarkey et al., 1995). Many of the ligands for receptors of this class are known to act as growth factors. As such, enormous effort has been directed to examining the details of these kinase cascades. These studies have been aimed at understanding how such agents induce mitogenesis and to address whether pharmacological interference in these cascades might represent an effective anti-cancer strategy.

For a considerable period of time studies on G protein-linked and tyrosine kinase-linked signalling cascades were essentially mutually exclusive. However, in recent times this has changed markedly with the appreciation that, at least in certain settings, G protein-coupled receptors have the capacity to regulate the activity of key kinases long appreciated to be modified by receptors which display intrinsic tyrosine kinase activity (Malarkey et al., 1995).

Signalling functions of lysophosphatidic acid

Lysophosphatidic acid (LPA) is a simple bioactive glycerophospholipid which is rapidly released from activated platelets and other cells (see Durieux and Lynch, 1993 and Moolenaar, 1995 for reviews). Although it has a diverse range of biological activities, one central feature is that it acts as a mitogen in many systems. This has particularly been studied in a number of fibroblast cell lines. LPA is able to cause enhanced GTP-loading of $p21^{ras}$ (Van Corven et al., 1993; Howe et al., 1993; Cook et al., 1993; Hordijk et al., 1994a) and enhanced phosphorylation and increased activity of the extracellularly regulated kinases of the MAP kinase family. These effects are also readily observed following stimulation of receptors with intrinsic tyrosine kinase activity, although via a very distinct mechanism involving recognition of the activated tyrosine kinase receptors by enzymatic polypeptides containing SH2 binding domains and subsequent transfer of information via adaptor proteins able to bind to protein SH3 domains (Malarkey et al., 1995).

In contrast to ligands such as epidermal growth factor (EGF) and platelet derived growth factor (PDGF), such effects of LPA are attenuated by pre treatment of cells with pertussis toxin (see van Corven et al., 1993, for example). Such observations imply a role for a G_i-like G-protein(s) in the function of this lipid and direct activation of G_i-like G proteins in membranes of Rat-1 fibroblast derived cell lines have been recorded (Carr et al., 1994). However, a seven transmembrane element receptor for LPA has not yet been purified nor has a relevant cDNA clone been isolated. A putative receptor protein has, however, been identified in ligand binding studies (Thomson et al., 1994) as has a 38–40 kDa LPA binding protein in cross-linking experiments (Van der Bend et al., 1992) but demonstration that this is a true G protein-coupled receptor is still lacking. Furthermore, the lack of a suitable pharmacology at this putative receptor means that it is not highly suitable for detailed studies on the mechanisms of G protein-linked receptor regulation of this cascade.

THE α_{2A}-ADRENOCEPTOR

Expression of the α_{2A}-adrenoceptor in Rat-1 fibroblasts

As a means to study the specificity or promiscuity of G protein coupling of the human α_{2A}-adrenoceptor we transfected Rat-1 fibroblasts with genomic DNA encoding this polypeptide. A clone, termed 1C, expressed some 3 pmol/mg membrane protein of this receptor based on the specific binding of the α_2-adrenoceptor antagonist [^{3}H]yohimbine. This receptor was clearly coupled to the cellular G protein population as the α_2-adrenoceptor agonist UK14304 caused stimulation of high affinity GTPase activity and inhibition of forskolin-amplified adenylyl cyclase (Milligan et al., 1991). As anticipated, both of these effects were abolished by pre-treatment of the cell with pertussis toxin. Detailed analysis indicated that the receptor was able to activate both of the pertussis toxin-sensitive G proteins G_i2 and G_i3 (Milligan et al., 1991). Others had noted previously that heterologous expression of this receptor could result in stimulation of inositol phosphate production (Cotecchia et al., 1990). Although we could occasionally record weak stimulation of a phosphoinositidase C activity in response to an α_2-adrenoceptor agonist, the response was trivial compared to the degree of activation produced by endothelin 1 acting on the endogenously expressed endothelin ET_A receptor (MacNulty et al., 1992). A far more robust lipid mediator response was noted when examining α_2-adrenoceptor agonist stimulation of a phosphatidylcholine-directed phospholipase D (PLD) activity. A similar activity was also stimulated in response to endothelin 1. However, whereas both the stimulation of inositol phosphate generation and of PLD activity in response to endothelin 1 were unaffected by pertussis toxin treatment of the cells, in parallel experiments such a treatment blocked the response to UK14304 (MacNulty et al., 1992). These cells thus provided a potentially useful model to examine signal transduction from G_i-linked receptors to MAP kinase.

When expressed in Rat-1 fibroblasts the α_{2A}-adrenoceptor causes GTP-loading of $p21^{ras}$ and activation of MAP kinases

Based on the ability of LPA to cause enhanced levels of GTP-loading of $p21^{ras}$ in Rat-1 fibroblasts in a pertussis toxin-sensitive fashion, Alblas et al. (1993) examined whether a similar effect would be observed upon agonist activation of the α_{2A}-adrenoceptor in clone 1C cells. In serum deprived cells, both UK14304 and EGF were able to cause a rapid increase in the proportion of cellular $p21^{ras}$ loaded with GTP. EGF but not UK14304 was able to produce a similar effect in parental Rat-1 fibroblasts. In clone 1C cells, the effect of UK14304 on GTP loading of $p21^{ras}$ was transient. Maximal levels were obtained within 2–5 minutes and decay to basal levels occurred over the next 10–20 minutes. As anticipated for effects produced via the α_{2A}-adrenoceptor, UK14304 stimulation of $p21^{ras}$ loading was attenuated by prior exposure of clone 1C cells to pertussis toxin while the effect of EGF was unaltered (Alblas et al., 1993). UK14304-stimulated $p21^{ras}$ loading in clone 1C cells was not prevented by the presence of the membrane permeant and phosphodiesterase resistant analogue of cAMP, 8-bromo-cAMP. Such a result tends to indicate that the effect of UK14304 on the fraction of cellular $p21^{ras}$ loaded with GTP was not a direct reflection of the ability of the α_2-adrenoceptor to cause inhibition of adenylyl cyclase and thus reduce cellular cAMP levels. As endothelin 1 is not able to promote GTP loading of $p21^{ras}$ in Rat-1 cells and clones derived there from (van Corven et al., 1993) it also appeared unlikely that the effects of the α_{2A}-adrenoceptor on this would be secondary to activation of phospholipases of the C or D classes.

Challenge of clone 1C cells with UK14304 led to the activation of both $p42^{MAPK}$ and $p44^{MAPK}$ which are central to the mitogenic process. This was determined by directly measuring the ability of the immunoprecipitated MAP kinases to increase the degree of

TABLE 1
Activation of MAP kinase activity in Rat-1 fibroblasts transfected to express the α_{2A}-adrenoceptor

Ligand	Rat-1 clone 1C cells	Rat-1 cells
	(fold stimulation over basal)	
PDGF (10 ng/ml)	2.9 ± 0.8	11.6
EGF (10 nM)	2.0 ± 0.5	6.3 ± 2.8
LPA (10 µM)	1.5 ± 0.3	5.9
UK14304 (1 µM)	2.3 ± 0.3	

MAP kinase activity was measured in immunoprecipitates by the phosphorylation of myelin basic protein added as substrate. The stimulation of activity observed by ligands at the endogenously expressed receptors (PDGF, EGF, LPA) was substantially lower in clone 1C compared to the parental cells due, at least in part, to a greater basal MAP kinase activity in these cells.

phosphorylation of myelin basic protein provided as substrate (Anderson and Milligan, 1994) (Table 1) or more indirectly by the increased tyrosine phosphorylation and the 'gel-shift' of $p42^{MAPK}$ and $p44^{MAPK}$ which is associated with their activation (Alblas et al., 1993; Anderson and Milligan, 1994). As anticipated from the work of others (see Burgering et al., 1993; Cook and McCormick, 1993; Sevetson et al., 1993; Hordijk et al., 1994b, for example), elevation of cellular cAMP levels limited the ability of UK14304 to activate the MAP kinases. Addition of UK14304 to serum-deprived clone 1C cells also resulted in an increase in the incorporation of [^{3}H]thymidine, demonstrating the mitogenic potential of the α_2-adrenoceptor in this system (Alblas et al., 1993).

α_{2A}-adrenoceptors and the regulation of phosphoinositide-3 kinase activity?

Activation of the MAP kinases by UK14304 in clone 1C cells is partially inhibited by relatively high concentrations (100 nM) of the fungal metabolite wortmannin. This agent is routinely described as a specific inhibitor of the enzyme phosphatidylinositol-3 kinase (PI-3 kinase) (Ui et al., 1995). However, direct measurements of PI-3 kinase activity in clone 1C cells indicates that UK14304 is able to produce very limited stimulation of this activity, particularly when compared to a more classical activator of this enzymatic activity such as PDGF. However, addition of UK14304, to clone 1C cells does result in a pertussis toxin-sensitive activation of the ribosomal S6 kinase, $p70^{S6K}$, as assessed both by an increase in its ability to phosphorylate a peptide substrate (Table 2) and by a reduction in the mobility of this polypeptide in SDS-PAGE, which as in the case of the MAP kinases reflects phosphorylation of this polypeptide. This activation of $p70^{S6K}$ was attenuated by pre treatment with wortmannin. As it is generally thought that $p70^{S6K}$ is downstream of PI 3-kinase these effects appear to represent something of a dichotomy. Little is currently known about the stoichiometry of signal transmission between members of these kinase cascades and thus it may be that only a small increase in PI-3 kinase activity is sufficient to cause activation of $p70^{S6K}$. Recent studies based both on partial purification (Stephens et al., 1994) and the isolation of a cDNA encoding a G protein activated PI-3 kinase, p110γ, (Stoyanov et al., 1995) have provided a potential rationale for the effect of the α_2-adrenoceptor in these cells. However, it remains to be examined whether this isoform of PI-3 kinase, as opposed to the more conventional p110α and p110β isoforms which associate, via the SH2 domains of the 85 kDa regulatory subunit, with the activated PDGF

TABLE 2

Agonist activation of the α_{2A}-adrenoceptor in clone 1C cells causes a pertussis toxin-sensitive stimulation of the ribosomal S6 kinase, p70^{S6K}.

Ligand	Control (cpm)	Pertussis toxin treated (25ng/ml)
Basal	15081	10253
PDGF (10 ng/ml)	21559	14859
LPA (10 µM)	24360	8590
UK14304 (1 µM)	34850	9797

Activity of the ribosomal S6 kinase, p70^{S6K} was assessed in immunoprecipitates from clone 1C cells by phosphorylation of a synthetic peptide substrate. The cells which were either controls or had previously been exposed to pertussis toxin (25 ng/ml, 16 h) were challenged with a variety of receptor ligands for 10 min prior to lysis and immunoprecipitation. The effects of both UK14304 and LPA but not PDGF were eliminated in the pertussis toxin treated cells.

receptor and other phosphotyrosyl proteins, is expressed in cells other than those of haematopoietic origin. UK14304 stimulation of p70^{S6K} activity, but not that of the MAP kinases, was attenuated by treatment of the cells with rapamycin. The intracellular target for rapamycin, following binding to its intracellular receptor, FKBP12, is thought to be rapamycin associated protein (FRAP) (Brown et al., 1994) which has limited sequence identity with the mammalian PI-3 kinase isoforms.

Comparison of the regulation of MAP kinases by wild type and mutant α_{2A}-adrenoceptors

Mutation of Asp 79 to Asn (D79N) has been reported to result in a marked reduction in the ability of the expressed receptor to contact the cellular G protein population. This was evidenced by a reduction in agonist stimulation of high affinity GTPase activity and suppression of the sensitivity to guanine nucleotides of agonist binding. However, this mutation has been reported not to interfere with agonist-mediated inhibition of adenylyl cyclase (Surprenant et al., 1992, Ceresa and Limbird, 1994).

Following stable expression of cDNA species encoding the wild type or D79N mutant α_{2A}-adrenoceptor in Rat-1 fibroblasts, clones were isolated expressing similar levels of receptors. The ability of UK14304 to stimulate high affinity GTPase activity was substantially higher for the wild type receptor than for the D79N mutant at the same level of receptor expression. A similar profile was observed for agonist-induced, cholera toxin-catalysed [^{32}P]ADP-ribosylation of G$_i$-like G-proteins. Agonist treatment of clones expressing similar levels of the wild type and D79N forms of the α_{2A}-adrenoceptor resulted in the phosphorylation and activation of both p42^{MAPK} and p44^{MAPK}. Concentration-effect curves for phosphorylation of these kinases showed the effect of agonist to be considerably more potent for the wild type receptor. Maximally effective concentrations of agonist resulted in a more rapid activation of p42^{MAPK} and p44^{MAPK} and a longer period of maintenance of the phosphorylated forms of these proteins in clones expressing the wild type receptor. This was not a reflection of simple clonal variation as both the concentration and time course of activation of these kinases in response to LPA were identical. As such, although the coupling efficiency of the D79N mutant of the α_{2A}-adrenoceptor to the population of G$_i$-like G proteins is severely impaired, even limited activation of the cellular G$_i$-population is sufficient to result in a maximal activation of the MAP kinases.

CONCLUSIONS

A number of examples have now been reported in which agonist activation of G_i-linked receptors can lead to activation of members of the MAP kinase family. Transfection of the α_{2A}-adrenoceptor into Rat 1 fibroblasts, a cell system which has been widely used to examine the ability of LPA to regulate elements of this cascade and to act as a mitogen, has allowed questions relating to how the levels of expression of G_i-coupled receptors and their G protein coupling efficiency can regulate the effectiveness of the MAP kinase cascade. The powerful available pharmacology at this receptor will also allow a detailed examination of the stoichiometry of interaction of the elements of this signalling cascade. Activation of the MAP kinase cascade by the α_{2A}-adrenoceptor in this cell system will also allow further examination of the interfaces between G protein-coupled and tyrosine kinase-linked cell signalling systems.

ACKNOWLEDGEMENTS

Work in the authors' laboratories in this area is supported by the Biotechnology and Biological Sciences Research Council.

References

Alblas, J.E., van Corven, E.J., Hordijk, P.L., Milligan, G. and Moolenaar, W.H. (1993). G_i-mediated activation of the p21^{ras}-mitogen-activated protein kinase pathway by α_2-adrenergic receptors expressed in fibroblasts. *Journal of Biological Chemistry*, **268**, 22235–22238.

Anderson, N.G. and Milligan, G. (1994). Regulation of p42 and p44 MAP kinase isoforms in Rat-1 fibroblasts stably transfected with α_2 adrenoceptors. *Biochemical and Biophysical Research Communications*, **200**, 1529–1535.

Brown, E.J., Albers, M.W., Shin, T.B., Ichikawa, K., Keith, C.T., Lane, W.S. and Schreiber, S.L. (1994). A mammalian protein targeted by G1-arresting rapamycin-receptor complex. *Nature*, **369**, 756–758.

Burgering, B.T.M., Pronk, G.J., van Weeren, P.C., Chardin, P. and Bos, J.L. (1993). cAMP antagonises p21^{ras}-directed activation of extracellular signal-regulated kinase 2 and phosphorylation of mSos nucleotide exchange factor. *EMBO J.*, **12**, 4211–4220.

Carr, C., Grassie, M. and Milligan, G. (1994). Stimulation of high affinity GTPase activity and cholera toxin-catalysed [^{32}P]ADP-ribosylation of G_i by lysophosphatidic acid (LPA) in wild-type and α_2C10 adrenoceptor-transfected Rat 1 fibroblasts. *Biochemical Journal*, **298**, 493–497.

Ceresa, B.P. and Limbird, L.E. (1994). Mutation of an aspartate residue highly conserved among G-protein-coupled receptors results in non-reciprocal disruption of α_2-adrenergic receptor-G-protein interactions. A negative charge at amino acid residue 79 forecasts α_{2A}-adrenergic receptor sensitivity to allosteric modulation by monovalent cations and fully effective receptor/G-protein coupling. *Journal of Biological Chemistry*, **269**, 29557–29564.

Cook, S.J. and McCormick, F. (1993). Inhibition by cAMP of Ras-dependent activation of Raf. *Science*, **262**, 1069–1072.

Cook, S.J., Rubinfield, B., Albert, I. and McCormick, F. (1993). RapV12 antagonises ras-dependent activation of ERK1 and ERK2 by LPA and EGF in Rat-1 fibroblasts. *EMBO Journal*, **12**, 3475–3485.

Durieux, M.E. and Lynch, K.R. (1993). Signalling properties of lysophosphatidic acid. *Trends in Pharmacological Sciences*, **14**, 249–254.

Hordijk, P.L., Verlaan, I., van Corven, E.J. and Moolenaar, W.H. (1994a). Protein tyrosine phosphorylation induced by lysophosphatidic acid in Rat-1 fibroblasts. Evidence that phosphorylation of MAP kinase is mediated by the G_i-p21^{ras} pathway. *Journal of Biological Chemistry*, **269**, 645–651.

Hordijk, P.L., Verlaan, I., Jalink, K., van Corven, E.J. and Moolenaar, W.H. (1994b). cAMP abrogates the p21^{ras}-mitogen-activated protein kinase pathway in fibroblasts. *Journal of Biological Chemistry*, **269**, 3534–3538.

Howe, L.R. and Marshall, C.J. (1993). Lysophosphatidic acid stimulates mitogen-activated protein kinase activation via a G-protein-coupled pathway requiring p21^{ras} and p74^{raf-1}. *Journal of Biological Chemistry*, **268**, 20717–20720.

MacNulty, E.E., McClue, S.J., Carr, I.C., Jess, T., Wakelam, M.J.O. and Milligan, G. (1992). α_2-C10 adrenergic receptors expressed in Rat 1 fibroblasts can regulate both adenylyl cyclase and phospholipase D-mediated hydrolysis of phosphatidylcholine by interacting with pertussis toxin-sensitive guanine nucleotide binding proteins. *Journal of Biological Chemistry*, **267**, 2149–2156.

Malarkey, K., Belham, C.M., Paul, A., Graham, A., McLees, A., Scott, P.H., *et al.* (1995). The regulation of tyrosine kinase signalling pathways by growth factor and G-protein-coupled receptors. *Biochemical Journal*, **309**, 361–375.

Milligan, G., Carr, C., Gould, G.W., Mullaney, I. and Lavan, B.E. (1991). Agonist-dependent, cholera toxin-catalysed ADP-ribosylation of pertussis toxin-sensitive G-proteins following transfection of the human α_2-C10 adrenergic receptor into Rat 1 fibroblasts. Evidence for the direct interaction of a single receptor with two pertussis toxin sensitive G-proteins, G_i2 and G_i3. *Journal of Biological Chemistry*, **266**, 6447–6455.

Moolenaar, W.H. (1995). Lysophosphatidic acid, a multifunctional phospholipid messenger. *Journal of Biological Chemistry*, **270**, 12949–12952.

Sevetson, B.R., Kong, X. and Lawrence J.C.Jr. (1993). Increasing cAMP attenuates activation of mitogen-activated protein kinase. *Proceedings of the National Academy of Sciences USA*, **90**, 10305–10309.

Stephens, L., Smrcka, A., Cooke, F.T., Jackson, T.R., Sternweis, P.C. and Hawkins, P.T. (1994). A novel phosphoinositide-3 kinase activity in myeloid-derived cells is activated by G protein $\beta\gamma$ subunits. *Cell*, **77**, 83–93.

Stoyanov, B., Volina, S., Hanck, T., Rubio, I., Loubtchenkov, M., Malek, D., *et al.* (1995). Cloning and characterisation of a G protein-activated human phosphoinositide-3 kinase. *Science*, **269**, 690–693.

Surprenant, A., Horstman, D.A., Akbarali, H. and Limbird, L.E. (1992). A point mutation of the α_2-adrenoceptor that blocks coupling to potassium but not calcium currents. *Science*, **257**, 977–980.

Thomson, F.J., Perkins, L., Ahern, D. and Clark, M. (1994). Identification and characterisation of a lysophosphatidic acid receptor. *Molecular Pharmacology*, **45**, 718–728.

Ui, M., Okada, T., Hazeki, K. and Hazeki, O. (1995). Wortmannin as a unique probe for an intracellular signalling protein, phosphoinositide 3-kinase. *Trends in Biochemical Sciences*, **20**, 303–307.

van Corven, E.J., Hordijk, P.L., Medema, R.H., Bos, J.L. and Moolenaar, W.H. (1993). Pertussis toxin-sensitive activation of p21[ras] by G protein-coupled receptor agonists in fibroblasts. *Proceedings of the National Academy of Sciences USA*, **90**, 1257–1261.

Van der Bend, R.L., Brunner, J., Jalink, K., van Corven, E.J., Moolenaar, W.H. and van Blitterswijk, W.J. (1992). Identification of a putative membrane receptor for the bioactive phospholipid lysophosphatidic acid. *EMBO Journal*, **11**, 2495–2501.

DIFFERENTIAL REGULATION OF α_2 ADRENERGIC RECEPTOR SUBTYPES BY NOREPINEPHRINE AND BUFFERS

D. ROSELYN CERUTIS, JEAN D. DEUPREE, DONALD A. HECK, SI-JIA ZHU, MYRON L. TOEWS and DAVID B. BYLUND

Department of Pharmacology, University of Nebraska Medical Center, 600 South 42nd Street, Omaha, NE 68198-6260

The physiological significance of multiple α_2 adrenergic receptor subtypes appears to be their differential regulation. The α_{2B} and α_{2C} subtypes are 25- to 200-fold more sensitive to downregulation by norepinephrine than is the α_{2A} subtype, depending on the species and cell type. In transfected cells, downregulation by norepinephrine results from an increased rate of degradation of the receptor protein, with no change in the rate of appearance of new receptors. Norepinephrine causes a two-fold increase in the degradation rate of the α_{2A} subtype and a four-fold increase in the degradation rate of the α_{2B} subtype. Both the K_d and the B_{max} values for the various subtypes observed in membrane binding studies can be significantly regulated by the assay buffer. The affinities of [^{3}H]rauwolscine for the α_{2A} and α_{2B} subtypes are several-fold higher in glycylglycine as compared to Tris. By contrast, the affinity for the α_{2C} subtype was 2- to 3-fold higher in NaPO$_4$ as compared to glycylglycine or Tris buffers. Surprisingly, the B_{max} values for all three subtypes are higher in NaPO$_4$ buffer as compared to glycylglycine. For [^{3}H]RX821002, the affinity for all four pharmacological subtypes is highest in NaPO$_4$ buffer. Compared to the other two buffers, the B_{max} values in NaPO$_4$ buffer are higher for the α_{2A}, α_{2B} and α_{2C} subtypes, but lower for the α_{2D} subtype. These results emphasize the need for careful evaluation of the assay conditions for each subtype in various experimental paradigms, and they suggest potentially significant differences in the molecular regulation of the subtypes by their environment.

KEY WORDS: α_2 adrenergic receptor, regulation, buffer, downregulation, synthesis, degradation, re-population kinetics

INTRODUCTION

Evidence for α_2 adrenergic receptor subtypes has come from binding and functional studies in various tissues and cell lines, and more recently from the molecular cloning of the genes for the receptors. On the basis of these studies, three genetic and four pharmacological α_2 adrenergic receptor subtypes have been defined. The four pharmacological subtypes are designated as α_{2A}, α_{2B}, α_{2C} and α_{2D}. Human platelets and the HT29 adenocarcinoma cell line are the prototypical tissues for the α_{2A} receptors (Bylund *et al.*, 1988). The α_{2B} subtype was defined in the NG-108 cell line and the neonatal rat lung (Bylund *et al.*, 1988). The α_{2C} was initially characterized in the OK cell line, which was derived from renal proximal tubule epithelium of the North American opossum (Murphy and Bylund, 1988); subsequently, opossum kidney was shown to express this receptor

---

Correspondence to: David B. Bylund, Department of Pharmacology, University of Nebraska Medical Center, 600 South 42nd Street, Omaha, NE 68198-6260, Telephone: (402)559-4788, Fax: (402)559-7495.

(Blaxall et al., 1991). The α_{2D} subtype was characterized in the bovine pineal gland (Simonneaux et al., 1991). It exhibits low affinity for rauwolscine, SKF104078, and BAM1303, a pharmacology which correlates well with the pharmacology of the α_2 receptor of the rat submandibular gland (Michel et al., 1989). The application of the techniques of molecular biology has yielded clones for the human α_{2A}, α_{2B}, and α_{2C} subtypes, designated C10, C2, and C4, respectively, for their chromosomal localizations (Kobilka et al., 1987; Lomasney et al., 1990; Weinshank et al., 1990; Regan et al., 1988). Orthologous genes or cDNAs have been cloned from rat, mouse, pig and opossum (see Bylund et al., 1994).

Although the physiological relevance of these receptor subtypes is still poorly understood, we and others have suggested that their differential regulation may be of fundamental importance. Exposure of these receptors to the endogenous neurotransmitter norepinephrine or other agonists causes desensitization and downregulation (see Toews et al., 1991). We have previously shown that the α_{2A} and α_{2C} subtypes undergo rapid functional desensitization following agonist preincubation (Jones et al., 1990). The desensitization characteristics are very similar for these two subtypes.

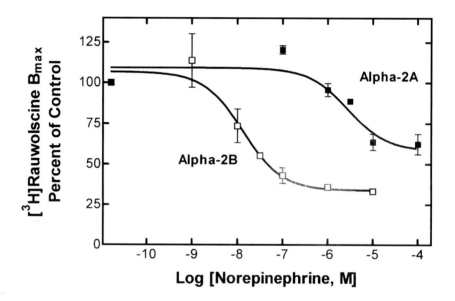

FIGURE 1 Norepinephrine concentration dependence of α_2 adrenergic receptor downregulation in CHO α_{2A} and CHO α_{2B} cells. Cells were pretreated with various concentrations of norepinephrine for 24 h, membranes isolated, and saturation binding experiments with [^{3}H]rauwolscine performed as previously described (Jones et al., 1991). The α_{2A} CHO cells have been previously characterized in our laboratory (Jones et al., 1991). The rat α_{2B} subtype (RNG) was obtained from Dr. Kevin Lynch, who kindly provided the 2.6 kb kidney cDNA in pGEM7Zf(+). The cDNA was then subcloned as a 2.6 kb HindIII/XbaI fragment into pRc/CMV (Invitrogen, San Diego, CA). Stable transfectants were established in CHO-K1 cells using the calcium phosphate precipitation technique. Stable transfectants were selected with 500 µg/ml G-418 (Genetecin; Life Technologies) and screened by intact cell binding (Toews et al., 1991) using [^{3}H]rauwolscine. The data shown are the means ± SEM from three experiments, expressed as percent of control. The receptor density (B_{max}) for control treatment groups for the α_{2A} was 1240 ± 90 fmol/mg protein and for the α_{2B} was 1250 ± 380 fmol/mg protein. The mean K_d values for all experiments were 0.82 ± 0.02 nM and 0.81 ± 0.08 nM for the α_{2A} and α_{2B} subtypes, respectively. The K_d values were not altered by the pretreatments.

TABLE 1
Downregulation of α_2 adrenergic receptor subtypes

	EC_{50}, nM	Extent, % of Control
α_{2A}		
HT29 cell	2,500	51
Transfected CHO	2,800	58
α_{2B}		
Transfected CHO	13	34
Transfected NIH-3T3	107	54
α_{2C}		
OK Cell	91	41

DIFFERENTIAL DOWNREGULATION OF α_2 ADRENERGIC RECEPTOR SUBTYPES

In order to investigate potential differences in the regulatory mechanisms of the α_2 subtypes, we compared the characteristics of downregulation following incubation with norepinephrine. CHO cells were stably transfected with the rat α_{2B} receptor and treated with 1 nM to 10 µM norepinephrine for 24 h. CHO cells transfected with the human α_{2A} (Fraser et al., 1989) were similarly treated with 0.1 µM to 100 µM norepinephrine and their downregulation was also characterized. When compared in the identical CHO cell background (Figure 1), the α_{2B} down-regulated to a greater extent than the α_{2A} subtype (35 vs 55 % of control; Table 1). More interestingly, the potency of norepinephrine in causing downregulation was significantly different for the two subtypes. The EC_{50} for norepinephrine was 13 nM for the α_{2B} subtype, whereas it was 2,800 nM for the α_{2A} subtype, a difference of over 200-fold. The low potency of norepinephrine in the CHO cells transfected with the α_{2A} subtype is consistent with our previous results, which found an EC_{50} of 3 µM for downregulation in these cells (Shreve et al., 1991).

Additional experiments have characterized the downregulation of the α_{2C} subtype and the α_{2A} and α_{2B} subtypes in other cellular backgrounds (Figure 2). Norepinephrine downregulated the α_{2C} receptor endogenously expressed in the OK cell with a potency of 91 nM, similar to its potency in down-regulating the α_{2B} receptor transfected into the CHO cell, but very different from the α_{2A} CHO cells. The α_{2B} receptor transfected into NIH-3T3 cells was potently down-regulated by norepinephrine similar to that in the CHO cell (Table 1). By contrast, norepinephrine had a low potency in down-regulating the α_{2A} receptor endogenously expressed in the HT29 cell, which was almost identical to its low potency in the transfected CHO cells.

The data summarized in Table 1 indicate a substantial difference between the α_{2A}, and the α_{2B} and α_{2C} subtypes in their sensitivity to downregulation by norepinephrine that appears to be intrinsic to the subtypes and largely independent of the cellular background. One simple explanation for this difference would be that norepinephrine has lower affinity for and potency at the α_{2A} subtype as compared to the other two subtypes. There is, however, little evidence to support such a hypothesis. In ^{3}H-antagonist binding assays, norepinephrine has a slightly higher affinity for the α_{2A} as compared to the α_{2B} subtype. In functional assays, norepinephrine behaves as a non-selective agonist at the three subtypes (Table 2). Thus the differential potency of norepinephrine in downregulating the

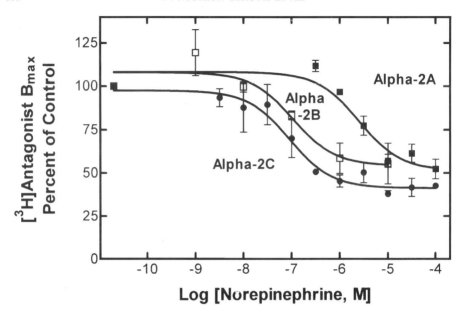

FIGURE 2 Norepinephrine concentration dependence of α_2 adrenergic receptor downregulation in HT29 cells (α_{2A}), NIH-3T3 cells transfected with the α_{2B}, and OK cells (α_{2C}). Cells were preincubated with various concentrations of norepinephrine for 24 h, membranes isolated, and saturation binding experiments with [^{3}H]rauwolscine (α_{2A} and α_{2C}) or [^{3}H]RX821002 (α_{2B}) performed as described previously (O'Rourke et al., 1994). The data are expressed as percent of control and represent the mean ± SEM. The HT29 and OK cell data are from Shreve et al. (1991). The NIH-3T3 cells transfected with the rat α_{2B} (RNG) receptor were kindly provided by Dr. Stephen Lanier. For the NIH-3T3 α_{2B} cells, the mean control B_{max} was 300 ± 34 fmol/mg protein (n = 4). The mean K_d value for all α_{2B} experiments was 1.95 ± 0.33 nM. The K_d values were not altered by the pretreatments.

α_2 subtypes appears to be an intrinsic property of the subtypes that is not directly related to their binding affinity or functional potency.

Eason and Liggett (1992) have shown that the human α_{2A} and α_{2B} subtypes transfected into CHO cells are down-regulated by a 24 h incubation with epinephrine. They only used one high concentration (100 µM) of epinephrine and found only a modest extent of downregulation (20 to 25 % of control). In addition they were unable to demonstrate downregulation of the human α_{2C} subtype. They attribute this lack of downregulation to

TABLE 2
Potency of norepinephrine at α_2 adrenergic receptor subtypes

	α_{2A}	α_{2B}	α_{2C}
	K_i or EC_{50}, nM		
^{3}H-Antagonist Binding	118	162	27
Adenylate Cyclase[a]	360	320	400
Cytosensor[b]	644	300	476
Down-regulation	2,800	13	91

[a]Data from Eason et al., 1994.
[b]Unpublished data from Roberts et al., 1994, Allergan.

the absence of a Cys residue in the COOH tail of the α_{2C} receptor equivalent to Cys-442 of the human α_{2A} receptor (Eason et al., 1994), which has been shown to be palmitoylated in the porcine α_{2A} subtype (Kennedy and Limbird, 1994). This conclusion was based on the loss of downregulation in a mutant (Cys -> Phe) α_{2A} receptor which would not be palmitoylated. Although this is an attractive hypothesis, the link between palmitoylation and downregulation is indirect at best. For example, neither the β_3 adrenergic receptor nor the 5-HT_{1A} serotonin receptor, which have the Cys residue, down-regulate in response to high agonist exposure (Eason et al., 1994). Conversely, the α_{2C} receptor of the OK cell lacks the appropriate Cys residue, but does indeed undergo downregulation in response to long-term norepinephrine exposure (Shreve et al., 1991; Pleus et al., 1993).

Whereas, it is possible that the discrepancy between our results (α_{2C} downregulation) and those of Liggett and co-workers relates to species differences (human vs opossum), recent experiments in our laboratory have provided an alternate explanation. In preliminary experiments, the downregulation of the opossum α_{2C} receptor transfected in the CHO cell shows a biphasic response to norepinephrine. At concentrations up to 0.3 µM, modest downregulation was evident, whereas at higher concentrations there was no downregulation but rather upregulation. Since Eason et al. used only a single high concentration of epinephrine, they may have missed downregulation occurring at lower concentrations. Upregulation of *transfected* receptors has also been observed recently with the α_1 adrenergic receptor. In some clones of CHO cells transfected with the α_{1B} receptor, preincubation with epinephrine results in a time- and concentration-dependent upregulation of receptor binding sites of approximately 2-fold (Zhu et al., 1996).

Thus, the differential downregulation of the α_2 subtypes, while subject to modulation by the cellular environment, is most likely a reflection of the intrinsic characteristics of each α_2 adrenergic receptor subtype.

MECHANISM OF AGONIST-INDUCED DOWNREGULATION

Since the density of a receptor at steady state is a function of its rates of appearance and disappearance, once a new steady state is reached following agonist pretreatment either the rate of appearance or disappearance or both must be altered. Many of the G protein-coupled receptors appear to fit a model in which the rate of receptor appearance is a zero order process and the rate of disappearance is a first order process. If r and k are the rate constants for appearance and disappearance, respectively, then the receptor level at steady state is equal to the ratio r/k (Neve and Molinoff, 1986). Since the characteristics of downregulation in the HT29 cells and in CHO cells transfected with the α_{2A} receptor are similar, we hypothesized that the changes in r and k would also be similar. In transfected cells the rate of synthesis is largely controlled by a strong viral promoter, and thus a change in the rate of degradation seems more likely to account for the lower density in the downregulated state. To determine r and k we used the method of repopulation kinetics after irreversible inactivation of the α_2 receptors by 10 µM 1-ethoxycarbonyl-2-ethoxy-1,2-dihydroquinoline (EEDQ). CHO α_{2A} receptor recovery in the presence of norepinephrine exhibited a two-fold increase in the rate of disappearance over receptors recovering in the absence of norepinephrine, corresponding to a decrease in the receptor half-life from 6.4 to 3.3 h. There was no change in the rate of receptor appearance between agonist treated and untreated cells. Thus, r/k decreased by 50%, consistent with the data shown in Figure 1 indicating an approximate 50% decrease in B_{max}. In the HT29 cell, we observed a slight (24%) decrease in the rate of appearance, but a larger (53%) increase in the rate of disappearance. In CHO cells transfected with the α_{2B} receptor, we found a dramatic increase in the rate of disappearance (half-life decreased from 3.5 to 0.9 h), with no decrease in the rate of appearance. This larger decrease in r/k is reflected in the greater extent of downregulation seen with the α_{2B} as compared to the α_{2A} (Figure 1).

These data suggest that the major difference between control and downregulated states is an increase in the rate of degradation. Thus, one can postulate that the mechanism of downregulation is simply the change in rate of degradation. If this is the case, then the time course of downregulation should be similar to the rate of disappearance in the down-regulated state. When we determined the time course for downregulation, however, we found that the $t_{1/2}$ for reaching the down-regulated state was not correlated with the $t_{1/2}$ for disappearance in the down-regulated state. For the CHO α_{2A}, HT29 and CHO α_{2B}, the $t_{1/2}$ values for downregulation were similar, 2.8, 3.0 and 3.7 hr, respectively, whereas the $t_{1/2}$ values of disappearance in the downregulated state were more variable, 3.3, 10.0 and 0.9 hr, respectively. Thus it appears that the mechanism of producing down-regulation may be different than the mechanism of maintaining the down-regulated state.

DIFFERENTIAL REGULATION OF α_2 SUBTYPES BY BUFFERS

Many factors have been shown to alter the binding characteristics of ligands for specific receptors, such as pH, ionic strength, osmolarity and metal ions; however, little attention has been given to the buffer in which the assay is conducted. It is generally assumed that all buffers are similar and that they do not significantly affect the binding characteristics of the receptor. Previously we noted that radiolabeled antagonists bound with higher affinity to α_{2A} adrenergic receptors when assayed in glycylglycine (Glygly) buffer than when assayed in Tris buffer (Jones et al., 1983), suggesting that buffers may produce enough of a conformational change in a receptor to alter the binding of antagonists to the receptor. We have now addressed three additional questions: 1) whether this phenomenon is unique to Glygly and Tris buffers; 2) whether all subtypes respond in the same manner to different buffers; and 3) whether buffers affect the binding of different radiolabeled antagonists similarly. Our approach was to examine the effects of Glygly, Tris, sodium phosphate ($NaPO_4$) and potassium phosphate (KPO_4) buffers on the binding parameters of [^{3}H]rauwolscine to the α_{2A}, α_{2B} and α_{2C} subtypes and on the parameters of [^{3}H]RX821002 binding to all four subtypes of the α_2 adrenergic receptor. Surprisingly, we found that both the affinities and binding site densities determined by radiolabeled antagonists could be altered by the assay buffer used, that not all subtypes responded in the same manner to a particular buffer and that different buffers had different effects on the binding characteristics of the radiolabeled antagonists.

A comparison of the affinities of [^{3}H]rauwolscine for the α_{2A}, α_{2B} and α_{2C} adrenergic receptors in the presence of the four buffers indicated that [^{3}H]rauwolscine has a significantly higher affinity (lower K_d) for the α_{2A} subtype in Glygly buffer compared with Tris, $NaPO_4$ or KPO_4 buffers (Figure 3). In contrast, the affinity of [^{3}H]rauwolscine for the α_{2B} receptor was not significantly different in Glygly, $NaPO_4$ or KPO_4 buffer, but

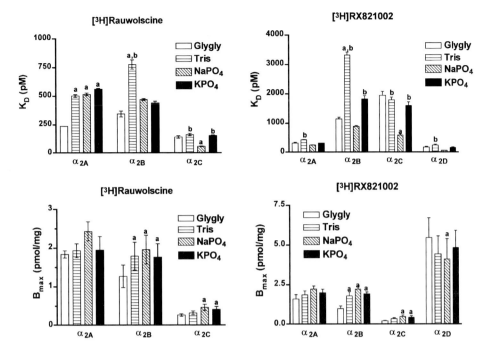

FIGURE 3 K_d and B_{max} values for [^{3}H]rauwolscine and [^{3}H]RX821002 binding to α_2 adrenergic receptor subtypes in various buffers. Saturation studies, measuring [^{3}H]rauwolscine and [^{3}H]RX821002 binding to the α_{2A}, α_{2B} and α_{2C} subtypes present in CHO cell membranes and the α_{2D} subtype in NIH-3T3 cells, were conducted as described previously (Bylund et al., 1988; Blaxall et al., 1991) in 25 mM concentrations of each buffer at pH 7.4. Experiments on [^{3}H]rauwolscine binding to the α_{2D} subtype were not conducted because of the very low affinity of [^{3}H]rauwolscine for this subtype. The K_d and B_{max} values were determined using the nonlinear regression analysis in the Prism computer program (GraphPad, San Diego, CA) with weighting by actual distances. The effects of all four buffers for a single subtype were determined in the same experiment. The graphs represent the geometric mean and SEM of the K_d values and the arithmetic means and SEM of the B_{max} values of three to four experiments. ANOVA and repeated measures analyses were used to compare the pK_d and B_{max} values, respectively. Tukey-Kramer post tests were used. [a]$p < 0.05$ compared with Glygly, [b]$p < 0.05$ compared with NaPO$_4$.

Tris produced a two-fold lower affinity for the ligand. [^{3}H]Rauwolscine had a 2.5- to 2.8-fold higher affinity for the α_{2C} receptor in NaPO$_4$ buffer compared with the other three buffers. The results of similar studies using [^{3}H]RX821002 as the radioligand indicated that the highest affinities were obtained with NaPO$_4$ buffer for all four subtypes (Figure 3). The lowest affinities for all four subtypes were obtained using Tris buffer. The largest difference in affinities (3.7-fold) was seen between [^{3}H]RX821002 binding to the α_{2B} subtype in NaPO$_4$ buffer compared with Tris buffer.

A comparison of B_{max} values for the two ligands showed that the B_{max} values for the α_{2A} subtype were similar in all four buffers (Figure 3). However, the B_{max} values for both ligands for the α_{2B} subtype were 1.4- to 2.2-fold lower in Glygly compared with the other three buffers. The B_{max} values for both antagonists for the α_{2C} subtype were 1.5- to 2.2-fold lower in Glygly than in either of the phosphate buffers. In contrast, [^{3}H]RX821002 had a 1.3-fold lower B_{max} for the α_{2D} subtype in NaPO$_4$ buffer compared with Glygly.

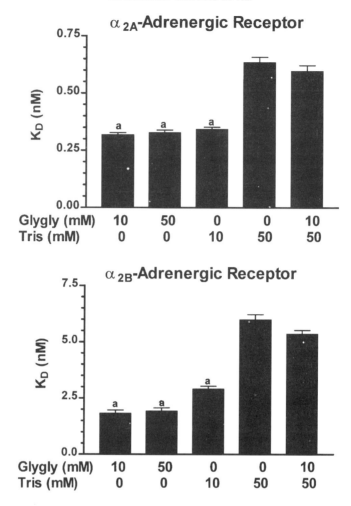

FIGURE 4 Effects of increasing concentrations of Glygly and Tris on the K_d and B_{max} for [^{3}H]RX821002 binding to the α_{2A} and α_{2B} subtypes. Binding of various concentrations of [^{3}H]RX821002 to both the α_{2A} and α_{2B} subtypes was determined in the presence of 10 and 50 mM Glygly and Tris, and 10 mM Glygly in the presence of 50 mM Tris as described in Figure 3. The geometric mean K_d values ± SEM from three to five experiments are presented. Statistical analyses were conducted as in Figure 3. [a]$p < 0.05$ compared with 50 mM Tris, and 50 mM Tris + 10 mM Glygly.

Since the lowest affinity of both [^{3}H]rauwolscine and [^{3}H]RX821002 for all subtypes studied was in Tris buffer, we wondered whether Tris was converting the receptor into a less favorable conformation or Glygly was converting the receptor into a more favorable conformation for ^{3}H-antagonist binding. To address this question we determined the effects of increasing concentrations of Glygly and Tris on the binding of [^{3}H]RX821002 (Figure 4) and [^{3}H]rauwolscine (not shown) to both the α_{2A} and α_{2B} subtypes. No significant changes in K_d for either subtype were seen when the concentration of Glygly was increased from 10 to 50 mM. However, the K_d values increased significantly when the concentration of Tris was increased from 10 to 50 mM for both subtypes.

These results suggest that buffers do not affect all subtypes of the α_2 adrenergic receptor similarly, nor are the affinities of [^{3}H]rauwolscine and [^{3}H]RX821002 for the different subtypes altered similarly by the four buffers. Although the binding site density for a specific subtype was altered differently by the different buffers, the B_{max} for a particular subtype was similar for both antagonists in a specific buffer. Thus, the buffers may be producing subtle conformational changes in the receptors to alter their ability to bind to various radiolabeled antagonists. This is most clearly seen by the fact that the affinities of both radioligands decreased as the concentration of Tris increased. These changes in binding parameters cannot be explained by changes in: a) ionic strength or osmolarity, since increasing the concentration of Glygly did not change the affinity of either radioligand for the receptor and the affinities in NaPO$_4$ and KPO$_4$ were quite different sometimes although the osmolarities of these two buffers were similar; or b) pH, since all buffers were adjusted to pH 7.4.

As a first step toward determining whether buffers alter the binding of agonists, we found that norepinephrine, in inhibiting [^{3}H]RX821002 binding, had a 3.2-fold higher affinity for the α_{2B} subtype in Tris buffer compared with NaPO$_4$ buffer. This suggests that Tris buffer may shift the receptor into a favorable conformation for agonists but not antagonists. If our further studies substantiate this fact, it would suggest that [^{3}H]RX821002 does not have equal affinity for all forms of the receptor and that [^{3}H]RX821002 may be an inverse agonist rather than a (pure) antagonist.

An obvious question is whether any one buffer can be used in ligand binding assays to give the highest affinity for the different radioligands for all four receptor subtypes. NaPO$_4$ buffer may be the most ideal for use in ligand binding studies since the affinity of [^{3}H]rauwolscine and [^{3}H]RX821002 is highest or as high as the other buffers for all subtypes except the α_{2A} subtype. The affinity of [^{3}H]rauwolscine for the α_{2A} subtype is two-fold lower in NaPO$_4$ buffer than in Glygly buffer. Tris buffer is not recommended for binding studies with either antagonist since it produced the lowest affinities for binding of either radioligand to any of the subtypes, and the decrease in affinity may be concentration-dependent.

References

Blaxall, H.S., Murphy, T.J., Baker, J.C., Ray, C. and Bylund, D.B. (1991). Characterization of the α_{2C} adrenergic receptor subtype in the opossum kidney and in the OK cell line. *J. Pharmacol. Exp. Ther.*, **259**, 323–329.

Bylund, D.B. (1988). Subtypes of α_2-adrenoceptors: pharmacological and molecular biological evidence converge. *Trends, Pharmacol. Sci.*, **9**, 356–361.

Bylund, D.B., Ray-Prenger, C. and Murphy, T.J. (1988). α_{2A} and α_{2B} adrenergic receptor subtypes: antagonist binding in tissues and cell lines containing only one subtype. *J. Pharmacol. Exp. Ther.*, **245**, 600–607.

Bylund, D.B., Eikenberg, D.C., Hieble, J.P., Langer, S.Z., Lefkowitz, R.J., Minneman, K.P., Molinoff, P.B., Ruffolo, R.R., Jr. and Trendelenburg, U. (1994). IV. International Union of Pharmacology Nomenclature of Adrenoceptors. *Pharmacol. Rev.*, **46**, 121–136.

Eason, M.G. and Liggett, S.B. (1992). Subtype-selective desensitization of α_2 adrenergic receptors. Different mechanisms control short and long-term agonist-promoted desensitization of α_2C10, α_2C4, and α_2C2. *J. Biol. Chem.*, **267**, 25473–25479.

Eason, M.G., Jacinto, M.T., Theiss, C.T. and Liggett, S.B. (1994). The palmitoylated cysteine of the cytoplasmic tail of α_{2A}-adrenergic receptors confers subtype-specific agonist-promoted downregulation. *Proc. Natl. Acad. Sci. USA*, **91**, 11178–11182.

Fraser, C.M., Arakawa, S., McCombie, W.R. and Venter, J.C. (1989). Cloning, sequence analysis, and permanent expression of a human α_2-adrenergic receptor in Chinese hamster ovary cells. *J. Biol. Chem.*, **264**, 11754–11761.

Jones, S.B., Bylund, D.B., Rieser, C.A., Shekim, W.O., Byer, J.A. and Carr, G.W. (1983). α_2-Adrenergic receptor binding in human platelets: alterations during the menstrual cycle. *Clin. Pharmacol. Ther.*, **34**, 90–96.

Jones, S.B., Leone, S.L. and Bylund, D.B. (1990). Desensitization of the α_2 adrenergic receptor in HT29 and opossum kidney cell lines. *J. Pharmacol. Exp. Ther.*, **254**, 294–300.

Jones, S.B., Halenda, S.P. and Bylund, D.B. (1991). α_2-Adrenergic receptor stimulation of phospholipase A_2 and of adenylate cyclase in transfected Chinese hamster ovary cells is mediated by different mechanisms. *Mol. Pharmacol.*, **39**, 239–245.

Kennedy, M.E. and Limbird, L.E. (1994). Palmitoylation of the α_{2A}-adrenergic receptor. Analysis of the sequence requirements for and the dynamic properties of α_{2A}-adrenergic receptor palmitoylation. *J. Biol. Chem.*, **269**, 31915–31922.

Kobilka, B.K., Matsui, H., Kobilka, T.S., Yang-Feng, T.L., Francke, U., Caron, M.G., Lefkowitz, R.J. and Regan, J.W. (1987). Cloning, sequencing, and expression of the gene encoding for the human platelet α_2-adrenergic receptor subtypes. *Science*, **238**, 650–656.

Lomasney, J.W., Lorenz, W., Allen, L.F., King, K., Regan, J.W., Yang-Feng, T.L., Caron, M.G. and Lefkowitz, R.J. (1990). Expansion of the α_2-adrenergic receptor family: Cloning and characterization of a human α_2-adrenergic receptor subtype, the gene for which is located on chromosome 2. *Proc. Natl. Acad. Sci. USA*, **87**, 5094–5098.

Michel, A.D., Loury, D.N. and Whiting, R.L. (1989). Differences between α_2-adrenoceptor in rat submaxillary gland and the α_{2A} and α_{2B} adrenoceptor subtypes. *Br. J. Pharmacol.*, **98**, 890–897.

Murphy, T.J. and Bylund, D.B. (1988). Characterization of α_2 adrenergic receptors in the OK cell, an opossum kidney cell line. *J. Pharmacol. Exp. Ther.*, **244**, 571–578.

Neve, K.A. and Molinoff, P.B. (1986). Turnover of β_1 and β_2 adrenergic receptors after down-regulation or irreversible blockade. *Mol. Pharm.*, **30**, 104–111.

O'Rourke, M.F., Blaxall, H.S., Iversen, L.J. and Bylund, D.B. (1994). Characterization of [^{3}H]RX821002 binding to α_2 adrenergic receptor subtypes. *J. Pharmacol. Exp. Ther.*, **268**, 1362–1367.

Pleus, R.C., Shreve, P.E., Toews, M.L. and Bylund, D.B. (1993). Down-regulation of α_2-adrenoceptor subtypes. *Eur. J. Pharmacol.*, **244**, 181–185.

Regan, J.W., Kobilka, T.S., Yang-Feng, T.L., Caron, M.G., Lefkowitz, R.J. and Kobilka, B.K. (1988). Cloning and expression of a human kidney cDNA for an α_2-adrenergic receptor subtype. *Proc. Natl. Acad. Sci. USA*, **85**, 6301–6305.

Shreve, P.E., Toews, M.L. and Bylund, D.B. (1991). α_{2A}- and α_{2C}-Adrenergic subtypes are differentially downregulated by norepinephrine. *Eur. J. Pharmacol.*, **207**, 275–276.

Simonneaux, V., Ebadi, M. and Bylund, D.B. (1991). Identification and characterization of α_{2D}-adrenergic receptors in bovine pineal gland. *Mol. Pharmacol.*, **40**, 235–241.

Toews, M.L., Shreve, P.E. and Bylund, D.B. (1991). Regulation of adrenergic receptors. In: Martinez, J.R., Edwards, B.S. and Seagrave, J.C. (Eds.) *Signaling Mechanisms in Secretory and Immune Cells*, pp.1–17. San Francisco: San Francisco Press.

Weinshank, R.L., Zgombick, J.M., Macchi, M., Adham, N., Lichtblau, H., Branchek, T.A. and Hartig, P.R. (1990). Cloning, expression, and pharmacological characterization of a human α_{2B}-adrenergic receptor. *Mol. Pharmacol.*, **35**, 681–688.

Zhu, S.J., Cerutis, D.R., Anderson, J.L. and Toews, M.L. (1996). Regulation of hamster α_{1B}-adrenoceptors expressed in Chinese hamster ovary cells. *Eur. J. Pharmacol.*, **299**, 205–212.

MOLECULAR BASIS OF α_2-ADRENERGIC RECEPTOR SUBTYPE REGULATION BY AGONIST

STEPHEN B. LIGGETT

Departments of Medicine and Pharmacology, University of Cincinnati College of Medicine, 231 Bethesda Avenue, Room 7511, Cincinnati, OH 45267-0564, USA

Some α_2-adrenergic receptor (α_2AR) subtypes undergo functional desensitization during continuous agonist exposure. Such desensitization is part of normal homeostasis and may also limit the effectiveness of therapeutic agonists. Based on studies using site-directed mutagenesis, recombinant expression, receptor phosphorylation and receptor function, it appears that short-term agonist-promoted desensitization is due to receptor phosphorylation by G-protein coupled receptor kinases (GRKs). The human α_{2A} and α_{2B} receptors undergo short-term desensitization via phosphorylation, while the α_{2C} receptor does not undergo desensitization. For the α_{2A}, the sites of GRK phosphorylation have been identified as four serines in the motif EESSSS which is in the third intracellular loop. There is also heterogeneity within the GRK family as to which kinase phosphorylates α_2AR. Co-expression studies indicate that desensitization of α_{2A} is due to phosphorylation by GRK2 (βARK) and GRK3 (βARK2) but not GRK5 or GRK6. Thus desensitization of α_2ARs by GRKs is both receptor subtype and kinase isoform specific.

Three human α_2-adrenergic receptor (α_2AR) subtypes have been cloned and have been designated α_2C10, α_2C4, and α_2C2 based on their human chromosomal localizations. These are equivalent to the pharmacologically defined α_{2A}, α_{2C}, and α_{2B} receptors, respectively. While all three of these subtypes bind endogenous catecholamines with about the same affinities, and display significant sequence homology throughout the transmembrane spanning domains, there are other regions of these receptors where there is divergence in the amino acid sequence. This structural heterogeneity suggests that there may be distinct subtype-specific functional properties of the different subtypes. To address this, we have transfected the three α_2AR subtypes individually into cells (such as CHW, CHO, COS-7, and HEK293) and have begun to examine receptor coupling and regulation. Indeed, we have found that the α_2AR subtypes do differ in terms of G-protein coupling (Eason *et al.*, 1992; Eason *et al.*, 1994a), short-term agonist-promoted desensitization by phosphorylation (Eason & Liggett, 1992), agonist-promoted sequestration (Eason & Liggett, 1992), and long-term agonist-promoted desensitization by downregulation (Eason & Liggett, 1992; Eason *et al.*, 1994b). In this chapter our progress in establishing a molecular basis for one of these processes, agonist-promoted desensitization by phosphorylation, will be discussed.

Correspondence to: Stephen B. Liggett, Depts. of Medicine and Pharmacology, University of Cincinnati College of Medicine, 231 Bethesda Avenue, Room 7511, Cincinnati, OH 45267-0564, USA. Tel: (513) 558-4831; Fax: (513) 558-0835

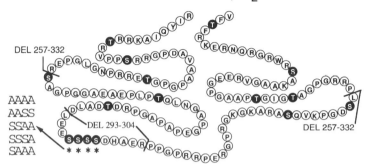

FIGURE 1 Primary amino acid sequence of the α_2C10 third intracellular loop. Indicated are various mutations as discussed in the text.

Desensitization, in general terms, is defined as a waning of a response despite the continued presence of a stimulus. Functional desensitization to continuous agonist exposure occurs with a number of G-protein coupled receptors (Liggett & Lefkowitz, 1993). The process is a homeostatic mechanism which occurs during normal physiologic conditions, may compensate or exacerbate responses during pathologic states, and may limit the therapeutic effectiveness of administered agonist.

α_2C10 UNDERGOES AGONIST PROMOTED DESENSITIZATION VIA βARK PHOSPHORYLATION

Our initial studies were aimed at determining if α_2AR (α_2C10) underwent agonist-promoted desensitization, and if so to determine the mechanism(s). At that time, only a few G-protein coupled receptors, such as rhodopsin, the β_2AR and the chick heart muscarinic receptor, were known to undergo agonist-promoted desensitization via receptor phosphorylation. However, it had been shown in a reconstituted phospholipid vesicle system using purified platelet α_{2A}AR and purified βAR kinase (βARK) that the receptor underwent phosphorylation by this kinase in an agonist-dependent manner (Benovic et al., 1987). So, phosphorylation was clearly a candidate mechanism. α_2C10 was permanently expressed in Chinese hamster fibroblasts (CHW-1102 cells), which do not endogenously express α_2AR. In adenylyl cyclase assays using membranes from these cells, α_2AR agonists mediated nearly 100% inhibition of 1.0 µM forskolin-stimulated activity. After brief exposure to agonist in culture, this functional coupling was depressed, in that the EC_{50} for epinephrine mediated inhibition of adenylyl cyclase was increased ~5-fold as compared to control cells (Liggett et al., 1992). The effect was concentration-dependent, with a maximal response being observed when cells were exposed to 10–100 µM epinephrine. This pattern of desensitization was identical when the cells were exposed to agonists for periods of time from 5 minutes up to 3 hours. After permeabilization of the cells, desensitization was blocked by the βARK inhibitor heparin, but not by PKA or PKC inhibitors.

Examination of the primary amino acid sequence of the α_2C10 (Figure 1), revealed a number of serines and threonines in the third intracellular loop which were potential

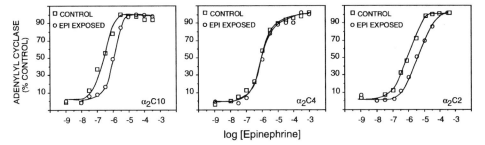

FIGURE 2 Differential agonist-promoted desensitization of α_2AR subtypes. CHO cells were exposed to 100 µM epinephrine for 30 minutes in culture, membranes were prepared, and then agonist-mediated inhibition of forskolin-stimulated adenylyl cyclase activities were determined. α_2C10 and α_2C2 displayed desensitization while α_2C4 did not. Shown are results from a single experiment. Modified from Eason & Liggett (1992).

βARK phosphorylation sites. We thus constructed a mutant receptor by site-directed mutagenesis which consisted of a deletion of the middle portion of the third intracellular loop, removing nine serines or threonines (Del 257-332, Figure 1). This receptor was expressed in CHW cells and in parallel studies with wild-type α_2C10 was found to not undergo agonist-promoted desensitization (Liggett et al., 1992). We then approached whether agonist-promoted desensitization of α_2C10 was accompanied by phosphorylation, and if so whether the mutant receptor failed to phosphorylate. Cells in monolayers were pre-incubated with ^{32}P-orthophosphate in media for two hours and some were treated with 100 µM epinephrine for various time periods. The cells were then washed with cold PBS, solubilized, and the receptor purified by immunoprecipitation with a subtype-specific antibody. α_2C10 was indeed found to undergo phosphorylation under these conditions while the mutant receptor did not (Liggett et al., 1992). Thus, there appeared to be a correlation between α_2C10 receptor phosphorylation and short-term agonist-promoted desensitization. Agonist-promoted sequestration of α_2C10 was unaffected by the mutation. Also, prolonged (24 hours) agonist exposure resulted in desensitization that was the same with the mutant as compared to the wild-type, pointing towards additional mechanism(s) involved in long-term desensitization (Liggett et al., 1992).

DIFFERENTIAL DESENSITIZATION AND PHOSPHORYLATION OF THE THREE α_2AR SUBTYPES

To address whether each of the α_2AR subtypes undergo short-term agonist-promoted desensitization, the human subtypes were permanently expressed in CHO cells and studied at expression levels of ~ 1 pmol/mg membrane protein (Eason & Liggett, 1992). (At this level of expression, the α_2AR in CHO cells primarily display G_i coupling. At higher expression levels a stimulatory response at high concentrations of agonist in the assay is observed). When the three receptors were studied in parallel, the effects of agonist (100 µM epinephrine) pre-exposure to cells for 30 minutes were clearly different between the three subtypes (Figure 2). For α_2C10, an ~ 5-fold increase in the EC_{50} was observed. At a submaximal concentration in the assay (1 µM), this shift in the dose response curve results in a ~ 70% desensitization of the response (Figure 2). In contrast, for α_2C4 the dose response curves for control and epinephrine pretreated cells were superimposable

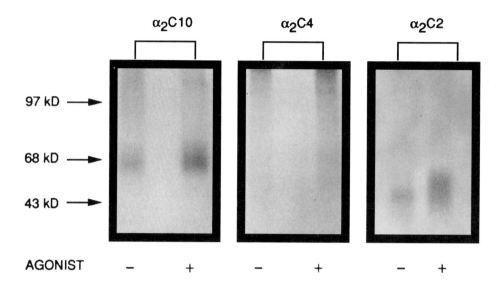

FIGURE 3 Differential agonist-promoted phosphorylation of α_2AR subtypes. Whole cell phosphorylation studies were carried out with the three human α_2AR subtypes expressed in CHO cells. α_2C10 and α_2C2 both displayed agonist promoted phosphorylation while α_2C4 did not.

(Eason & Liggett, 1992). No changes in the EC_{50}s or the maximal extents of inhibition were noted with α_2C4, despite the use of several different agonists, high concentrations, and exposure times up to 1 hour. For α_2C2, less desensitization as compared to α_2C10 was noted. This amounted to a ~2–3-fold increase in the EC_{50}. While this increase is relatively small, it is readily detectable and reproducible. This shift in the dose response curve resulted in ~30% desensitization of the response at 1 µM epinephrine in the assay (Eason & Liggett, 1992).

Based on work with α_2C10, we considered that α_2C10 and α_2C2 would undergo agonist-promoted phosphorylation, while α_2C4 would not. This indeed turned out to be the case. Whole-cell phosphorylation studies have been carried out with CHO cells expressing equal levels of each subtype as described above, and a typical result is shown in Figure 3. As can be seen, there is no agonist-promoted phosphorylation of α_2C4, while the other two subtypes do display such phosphorylation. Thus, the lack of agonist-promoted phosphorylation of α_2C4 is consistent with the lack of agonist-promoted functional desensitization. This issue has been explored by others as well. Kurose and Lefkowitz (1994) also found that α_2C10 and α_2C2 expressed in COS-7 cells underwent agonist-promoted phosphorylation in whole-cell phosphorylation studies while α_2C4 did not. In addition, in a reconstituted phospholipid vesicle system using purified α_2C4 and purified βARK, the receptor failed to display agonist-promoted phosphorylation.

The differential desensitization of the subtypes was not predicted based on the amino acid sequence of the third intracellular loops of these receptors. Each has a similar number of serines or threonines (α_2C10 = 16, α_2C4 = 20, α_2C2 = 20). We felt that in order to fully understand this subtype-specific phenomenon, a determination of the precise requirements for α_2AR phosphorylation by βARK was in order.

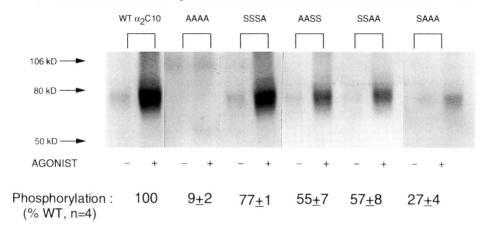

FIGURE 4 Agonist-promoted phosphorylation of mutated α_2C10 receptors. COS-7 cells were co-transfected with constructs encoding for the indicated mutant (see Figure 1) and βARK. Whole-cell phosphorylation studies were then carried out. Shown is an autoradiogram from a representative study and mean results from four studies. Modified from Eason et al. (1995).

FOUR SERINES ARE PHOSPHORYLATED IN α_2C10 DURING AGONIST-PROMOTED DESENSITIZATION

To delineate the requirements for βARK mediated phosphorylation of α_2C10, in recent studies we have constructed and studied six mutant receptors (Eason et al., 1995). Onorato et al. (1991), using small peptides in a reconstituted phospholipid vesicle system with purified βARK, had shown that acid residues preceeding serines markedly enhanced βARK mediated phosphorylation. This pointed towards the sequence EESSSS in the third intracellular loop of α_2C10 as a candidate phosphorylation site for βARK during agonist-promoted desensitization. To approach this, a small deletion mutant (Del 293-304, Figure 1) was constructed. This mutant receptor, when expressed in CHO cells, failed to undergo agonist-promoted functional desensitization, supporting the notion that this was the region of phosphorylation of the receptor. The following mutants, consisting of various substitutions of serines with alanines within the SSSS sequence (Figure 1) were then constructed: AAAA, AASS, SSAA, and subsequently SSSA and SAAA. Receptors were coexpressed in COS-7 cells with βARK and whole cell agonist-promoted phosphorylation studies were carried out as described above. The results are depicted in Figure 4. The AAAA mutant failed to phosphorylate. Removal of two serines, in mutants AASS or SSAA, resulted in a receptor that phosphorylated to a level which was 50% that of wild type α_2C10. This prompted construction of the SSSA and SAAA mutants, which were phosphorylated to levels of 75% and 25%, respectively, of wild-type receptors. Thus, removal of each serine decreased phosphorylation by 25%, consistent with all four serines being phosphorylated in the wild type and an implied stoichiometry of 4 moles of phosphate/mole of receptor. In functional desensitization experiments carried out in CHO cells, the AAAA mutant failed to desensitize as expected. However, the SSAA and AASS mutants, which both displayed partial agonist-promoted phosphorylation, also failed to undergo desensitization

(Eason et al., 1995). We have concluded from these results that "full" phosphorylation is required for βARK mediated agonist-promoted desensitization of α_2C10. The basis for this, presumably, is that the binding of βarrestin (which causes the uncoupling of the phosphorylated receptor from the G-protein) has precise conformational requirements.

From these results, and those of peptide based studies, it can be predicted that potential substrates for βARK include serines or threonines with amino-terminal aspartic or glutamic acids, and that the separation between the acid residues and the phosphoacceptor can be at least two residues. However, α_2C4 has at least one sequence (DESS) which would appear to be a candidate for βARK mediated phosphorylation. However, in the context of α_2C4, it does not appear to be so. This may be because it is located in a different position (21 residues more N-terminal) within the third intracellular loop of α_2C4 as compared to the EESSSS sequence in α_2C10. Also, this sequence is followed by a series of basic residues (arginines) which may alter the context sufficiently so that these sites are not favorable for βARK phosphorylation. In this regard we have also noted an extraordinary stretch of glutamic acid residues (EDEAEEEEEEEEEEEE) in the middle of the third intracellular loop of α_2C2. This is flanked on both sides by serines and threonines. We recently removed this acidic portion by either deletion of the residues or substitutions with glutamines (thereby conserving size but not charge). These mutated receptors failed to desensitize and only partially phosphorylated (Jewell-Motz & Liggett, 1995). Thus the surrounding milieu and the location of a potential phosphorylation site is critical to whether it is phosphorylated at all or whether such phosphorylation has a functional effect. This may explain a discrepancy in published reports concerning agonist-promoted desensitization of the α_{2C} receptor. As discussed earlier two groups have reported a lack of desensitization and phosphorylation of the human α_2C4 subtype. However, Jones et al. (1990) have reported agonist-promoted desensitization of the opossum receptor which appears to be the species homologue to α_2C4. The third intracellular loops of the two receptors, however, display distinct differences that may account for these results. Although not present in α_2C4, the opossum receptor has the sequence EESSTS in the middle of the loop (Blaxall et al., 1994). This is roughly in the same position as the highly homologous EESSSS sequence in α_2C10 which has been shown to be phosphorylated during agonist-promoted desensitization. Another striking difference between the human and opossum receptor is a stretch of histidines (17 out of 20 residues) in the third intracellular loop of the opossum receptor. The human receptor has only one histidine in the entire loop. The significance of this feature is not known but a difference in the conformation of the loops between the two receptors is likely.

GRKs DISPLAY SUBSTRATE SPECIFICITY FOR α_2ARs

We have also considered the possibility that α_2AR subtypes may serve as substrates for some GRKs but not others. Given that expression of the known kinases varies significantly in different cells and tissues, GRK substrate specificity may be the basis of cell-type heterogeneity of α_2AR desensitization. That the GRKs may have different requirements for binding or phosphorylation of G protein coupled receptors is suggested by the significant lack of homology in primary structure in several key domains of the GRKs which have been cloned to date (Inglese et al., 1993).

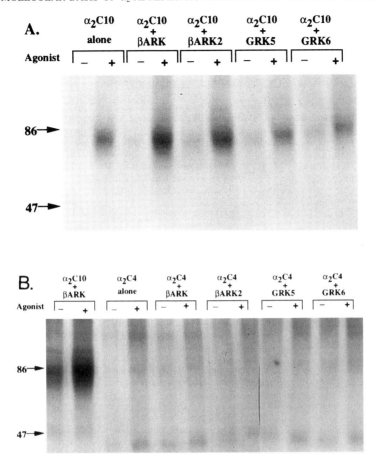

FIGURE 5 GRK substrate specificity for phosphorylating α_2AR subtypes. COS-7 cells were transfected with the indicated receptors and GRKs. βARK and βARK2 enhanced agonist-promoted phosphorylation of α_2C10, while GRK5 and GRK6 had no effect (Panel A). For α_2C4, no agonist-promoted phosphorylation was observed. This lack of phosphorylation was not affected by co-expression of any GRK (Panel B).

To approach this question, we transiently co-expressed the different GRKs with either α_2C10 or α_2C4 in COS-7 cells and carried out intact cell phosphorylation studies as described earlier. For these experiments GRK2 (βARK), GRK3 (βARK2), GRK5 and GRK6 were studied. These are expressed in a number of tissues in the CNS, cardiovascular, pulmonary and endocrine systems. (GRK4 expression is limited to the testis.) The basis for the approach resides in the observation that with marked overexpression of a receptor, the amount of endogenous GRK is insufficient to phosphorylate all the receptors. With co-overexpression of a given GRK, an enhanced phosphorylation signal would be expected *if* that particular GRK indeed phosphorylates the receptor. For these studies Western blots were utilized to ensure that GRK overexpression occurred and radioligand binding was used to monitor α_2AR expression.

Typical results from these studies (Jewell-Motz and Liggett, 1996) are shown in Figure 5. For α_2C10, co-expression with βARK and βARK2 resulted in an enhanced phosphorylation. In contrast, GRK5 and GRK6 co-expression did not alter agonist-promoted phosphorylation. With α_2C4, agonist-promoted phosphorylation was not noted in cells which had not been co-transfected with a GRK. This was consistent with the functional phosphorylation studies carried out in CHO cells. Co-expression of βARK, βARK2, GRK5 or GRK6 had no effect on α_2C4 phosphorylation. As is shown, α_2C10 was always included as a positive control in these studies. Functional studies of desensitization of α_2C10 and α_2C4 with GRK co-expression in HEK293 cells are consistent with these phosphorylation studies (Jewell-Motz and Liggett, 1996).

CONCLUSION

The α_2C10 subtype undergoes short-term agonist promoted desensitization via receptor phosphorylation. A large body of evidence has accumulated which indicates that phosphorylation of α_2C10 under these conditions is mediated by a GRK. This includes: the concentration and time dependency of desensitization, inhibition of desensitization by heparin, phosphorylation of purified α_2C10 by βARK in phospholipid vesicles, phosphorylation by βARK of a peptide derived from the third intracellular loop of α_2C10 the correlation of functional desensitization and *in vivo* receptor phosphorylation in whole cells, enhanced receptor phosphorylation in cells overexpressing βARK, and loss of desensitization and phosphorylation in receptors in which predicted βARK sites were mutated. The specific sites in α_2C10 that are phosphorylated by βARK have been determined.

Less is known about the regulatory mechanisms associated with agonist exposure to α_2C4 and α_2C2. Under identical conditions α_2C10 and α_2C2 undergo short-term agonist promoted functional desensitization which is accompanied by phosphorylation while α_2C4 does not. Thus, human α_2AR desensitization appears to be subtype-specific. In addition to GRK2 (βARK), GRK3 (βARK2) also phosphorylates α_2C10 in an agonist-dependent manner. On the other hand, GRK5 and GRK6 do not phosphorylate α_2C10. α_2C4 does not appear to be phosphorylated by any of the aforementioned kinases. Thus, it appears that α_2AR desensitization is also kinase-specific.

ACKNOWLEDGEMENTS

The work described here is funded by grants HL53436, HL41496, HL07382 and HL45967 from the National Institutes of Health and a gift from Procter and Gamble Pharmaceuticals. The substantial contributions of Drs. Margaret Eason and Elizabeth Jewell-Motz to these projects is most appreciated.

References

Benovic, J.L., Regan, J.W., Matsui, H., Mayor, F., Cotecchi, S., Leeb-Lundberg, L., *et al.* (1987). Agonist-dependent phosphorylation of the α_2-adrenergic receptor by the β-adrenergic kinase. *J. Biol. Chem.*, **262**, 17251–17253.

Blaxall, H.S., Cerutis, D.R., Hass, N.A., Iversen, L.J. and Bylund, D.B. (1994). Cloning and expression of the α_{2C}-adrenergic receptor from the OK cell line. *Mol. Pharmacol.*, **45**, 171–176.

Eason, M.G., Kurose, H., Holt, B.D., Raymond, J.R. and Liggett, S.B. (1992). Simultaneous coupling of α_2-adrenergic receptors to two G-proteins with opposing effects: Subtype-selective coupling of α_2C10, α_2C4 and α_2C2 adrenergic receptors to G_i and G_s. *J. Biol. Chem.*, **267**, 15795–15801.

Eason, M.G., Jacinto, M.T. and Liggett, S.B. (1994a). Contribution of ligand structure to activation of α_2AR subtype coupling to G_s. *Mol. Pharmacol.*, **45**, 696–702.

Eason, M.G., Jacinto, M.T., Theiss, C.T. and Liggett, S.B. (1994b). The palmitoylated cysteine of the cytoplasmic tail of α_{2A}-adrenergic receptors confers subtype-specific agonist-promoted downregulation. *Proc. Natl. Acad. Sci. USA*, **91**, 11178–11182.

Eason, M.G., Moreira, S.P. and Liggett, S.B. (1995). Four consecutive serines in the third intracellular loop are the sites for βARK-mediated phosphorylation and desensitization of the α_{2A}-adrenergic receptor. *J. Biol. Chem.*, **270**, 4681–4688.

Eason, M.G. and Liggett, S.B. (1992). Subtype-selective desensitization of α_2-adrenergic receptors: Different mechanisms control short and long-term agonist-promoted desensitization of α_2C10, α_2C4 and α_2C2. *J. Biol. Chem.*, **267**, 25473–25479.

Inglese, J., Freedman, N.J., Koch, W.J. and Lefkowitz, R.J. (1993). Structure and mechanism of the G protein-coupled receptor kinases. *J. Biol. Chem.*, **268**, 23735–23738.

Jewell-Motz, E.A. and Liggett, S.B. (1995). An acidic motif within the third intracellular loop of the α_2C2 adrenergic receptor is required for agonist-promoted phosphorylation and desensitization. *Biochem.*, **34**, 11946–11953.

Jewell-Motz, E.A. and Liggett, S.B. (1996). G protein coupled receptor kinase specificity for phosphorylation and desensitization of α_2-adrenergic receptor subtypes. *J. Biol. Chem.*, In press.

Jones, S.B., Leone, S.L. and Bylund, D.B. (1990). Desensitization of the α_2-adrenergic receptor in HT29 and opossum kidney cell lines. *J. Pharmacol. Exp. Ther.*, **254**, 294–300.

Kurose, H. and Lefkowitz, R.J. (1994). Differential desensitization and phosphorylation of three cloned and transfected α_2-adrenergic receptor subtypes. *J. Biol. Chem.*, **269**, 10093–10099.

Liggett, S.B., Ostrowski, J., Chestnut, L.C., Kurose, H., Raymond, J.R., Caron, M.G., *et al.* (1992). Sites in the third intracellular loop of the α_{2A}-adrenergic receptor confer short-term agonist-promoted desensitization: evidence for a receptor kinase-mediated mechanism. *J. Biol. Chem.*, **267**, 4740–4746.

Liggett, S.B. and Lefkowitz, R.J. (1993). In *Regulation of cellular signal transduction pathways by desensitization and amplification* edited by D. Sibley and M. Houslay, pp. 71–97, London: John Wiley & Sons.

Onorato, J.J., Palczewski, K., Regan, J.W., Caron, M.G., Lefkowitz, R.J. and Benovic, J.L. (1991). The role of acidic amino acids in peptide substrates of the β-adrenergic receptor kinase and rhodopsin kinase. *Biochem.*, **30**, 5118–5125.

α_2-ADRENERGIC RECEPTOR SUBTYPES DISPLAY DIFFERENT TRAFFICKING ITINERARIES IN MADIN-DARBY CANINE KIDNEY II CELLS

MAGDALENA WOZNIAK and LEE E. LIMBIRD

Department of Pharmacology, Vanderbilt University School of Medicine, Nashville, TN 37232-6600, USA

Subtypes of the α_2-adrenergic receptor (α_{2A}, α_{2B}, and α_{2C}AR) couple to the pertussis toxin-sensitive G_i and G_o subpopulations of G proteins and mediate a variety of physiological effects, including transcellular ion transport in renal epithelial cells. Previous studies have revealed that the α_{2A}AR is directly targeted to the lateral subdomain of polarized, cultured Madin-Darby Canine Kidney (MDCK) II cells *via* a pertussis toxin-insensitive mechanism; mutational analysis suggests that regions in or near the bilayer are critical for direct targeting of the α_{2A}AR whereas endofacial domains contribute to α_{2A}AR retention in the lateral surface. Unique structural characteristics in the α_{2A}, α_{2B} and α_{2C}AR subtypes led us to evaluate their targeting and retention in polarized MDCK II cells. Although the α_{2B}AR, like the α_{2A}AR, achieves 85–95% basolateral localization at steady-state, this polarization occurs after initial random insertion into both apical and basolateral surfaces, followed by selective retention and/or rerouting to the lateral subdomain; pulse-chase studies confirm that the half-life of the α_{2B}AR on the apical surface is markedly shorter than on the basolateral surface of MDCK II cells. The α_{2C}AR also is expressed on the lateral subdomain and, like the α_{2A}AR, achieves its localization *via* direct delivery to the basolateral surface. However, the α_{2C}AR also demonstrates significant intracellular localization, which others have interpreted as a precursor receptor pool. The findings indicate that the α_{2A}, α_{2B} and α_{2C}AR subtypes, which possess highly homologous structures, nonetheless manifest distinct delivery mechanisms, thus providing useful models for revealing structural motifs that confer unique trafficking information.

KEY WORDS: MDCK, α_2 adrenergic receptor subtypes, basolateral, trafficking, renal epithelial cells, localization

α_2-Adrenergic receptors (α_2-ARs) participate in multiple physiological effects *via* pertussis toxin-sensitive signal transduction pathways, including inhibition of adenylyl cyclase, activation of receptor-operated K^+ channels and suppression of voltage-gated Ca^{++} channels (Limbird, 1989). Since many cell types expressing the α_2-AR, including kidney and intestinal epithelium, vascular endothelium, and neurons are polarized, the physiological functions mediated by the α_2-AR rely on the presence of this receptor on the appropriate surface of polarized cells.

Precise targeting of molecules is crucial in many physiological processes. Several known diseases result from mislocalization of molecules; for example, in cystic fibrosis, a point mutation in the cystic fibrosis chloride transport regulator (CFTR) leads to its retention in the endoplasmic reticulum and consequently to a lack of appropriate Cl^- and H_2O transport at the apical surface of epithelial cells (Welsh *et al.*, 1992). An example

---

Correspondence: Magdalena Wozniak, Dept. of Pharmacology, Vanderbilt University School of Medicine, Nashville, TN 37232-6600, USA. Tel: 615-343-3533, Fax: 615-343-1084

that illustrates the physiological importance of correct targeting of a G protein-coupled receptor is retinitis pigmentosa, where certain mutations in rhodopsin, a G protein-coupled receptor, result in its retention in the endoplasmic reticulum, ultimately leading to blindness (Sung et al., 1993).

Functional evidence existing in the literature suggests that a high degree of specificity in receptor-G protein coupling may result from a precise co-localization of receptor molecules and G proteins with which they interact in the plasma membrane. For instance, the α subunit of G_i was shown to localize non-homogeneously to patches in the plasma membrane (Muntz et al., 1992). In addition, a lack of cross-talk in receptor-G protein coupling has been observed among muscarinic, α_2-AR and δ-opioid receptors in NG-108-15 cells, based on evaluation of guanine-nucleotide regulation of agonist binding, despite the fact that these three receptor populations all couple to G_i/G_o GTP-binding proteins (Graeser and Neubig, 1993). This apparent lack of cross-talk could be explained by a spatial separation of receptors and the G proteins with which they interact. It also has been shown by antisense "knockouts" that although muscarinic and somatostatin receptors both couple to the same class of G proteins — G_o, and consequently modulate the same effector system — suppression of voltage-sensitive Ca^{++} currents, each receptor type nonetheless uses a different combination of G_o subunits α, β and γ to modulate channel function (Kleuss et al., 1991; 1992; 1993). Such specificity cannot be explained by known receptor-G protein affinities or cell-specific subunit expression, and may result from localization of receptor populations and their respective G_o subunits within discrete microdomains in the plasma membrane.

The evidence for a precise localization of various receptor and G protein populations in the plasma membrane encouraged us to study the targeting of a G protein — coupled receptor. The existence of three subtypes of α_2-AR that share structural similarity but also possess distinct regions of sequence diversity as well as the role of endogenous α_2-AR in regulating the physiological processes in polarized cells provided motivation to study the targeting and stabilization of α_2-AR subtypes in polarized cells. Sequences in the seven transmembrane-spanning regions are highly homologous among the three α_2-AR subtypes, whereas sequences in the third cytoplasmic loop, amino- and carboxy-termini of the α_2-AR subtypes are quite distinct (Guyer et al., 1990; Kobilka et al., 1987; Lanier et al., 1991; Zeng et al., 1990). In addition, the post-translational modification differs among the three α_2-AR subtypes. For instance, the α_{2A}-AR has two consensus sites for N-linked glycosylation in its amino-terminal sequence and is acylated at a cysteine residue in position 442 in its carboxy-terminus; the α_{2B}-AR is not glycosylated but possesses a cysteine residue in the carboxy-terminus analogous to that acylated in the α_{2A}-AR; the α_{2C}-AR, in turn, is glycosylated but lacks the cysteine residue utilized for acylation in the α_{2A}-AR in its carboxy-terminus (Guyer et al., 1990; Kobilka et al., 1987; Lanier et al., 1991; Zeng et al., 1990).

In order to determine if the α_2-AR subtypes are differentially localized in polarized cells, each subtype of the α_2-AR, both wild type- and hemagglutinin-tagged, was stably expressed in the Madin-Darby canine kidney cell line (MDCK II), a polarized renal epithelial cell line. Permanent clonal cell lines were grown on Transwells for 7 days in order to polarize the α_2-AR-expressing MDCK II cells functionally and morphologically (Keefer and Limbird, 1993; Wozniak and Limbird, 1996). The integrity of the MDCK II cell monolayer was assessed before each experiment, based on lack of permeability to [^{3}H]-inulin (Keefer and Limbird, 1993).

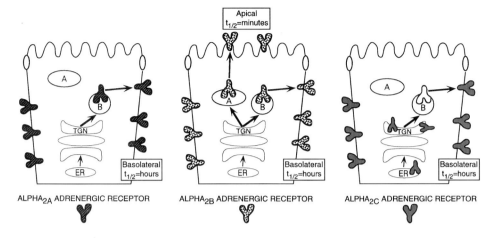

FIGURE 1 Schematic representation of differential targeting itineraries of the three α_2AR subtypes in MDCK II cells. This schematic diagram compares the targeting mechanisms and steady-state localization of the α_{2A}AR, α_{2B}AR and α_{2C}AR subtypes. Similarly to the α_{2A}AR, the α_{2C}AR is delivered directly to the basolateral membrane. The α_{2B}AR is inserted randomly into both cell surface domains and is retained preferentially on the basolateral surface. At steady-state, the α_{2A}AR and α_{2B}AR are detected exclusively on the lateral surface of MDCK II cells, whereas the α_{2C}AR is distributed between cell surface and intracellular compartments.

Figure 1 is a schematic diagram that summarizes the findings of these studies. The steady-state localization of each α_2-AR subtype on the apical *versus* basolateral domain of MDCK II cells was determined by biotinylating clonal cells on either the apical or the basolateral surface, followed by a covalent labeling of the α_2-AR in membrane extracts using an iodinated, photoactivatable α_2-AR-selective ligand [^{125}I]-Rau-AzPEC (Lanier *et al.*, 1986). Detergent-extracted, biotinylated receptor was then isolated by streptavidin-agarose chromatography. The amount of the α_2-AR in the streptavidin-agarose eluate detected following SDS-PAGE provides a quantitative assessment of the fraction of the α_2-AR present on the biotinylated surface (Keefer and Limbird, 1993). Our experiments indicate that the surface localization of all three subtypes in MDCK II cells is almost exclusively on the basolateral surface. Importantly, the localization of each α_2-AR subtype is not altered if the epitope tag is introduced into the amino- or carboxy-terminus of the receptor. This quantitative examination of the localization of each α_2-AR subtype using biochemical strategies was corroborated by immunocytochemical methods. Clonal cell lines expressing α_2-AR subtypes were immunostained using 12CA5 antibody directed against the hemagglutinin epitope tag. All three subtypes of α_2-AR were morphologically detected on the surface of MDCK II cells, and this surface expression was confined to the lateral subdomain of the basolateral membrane. Whereas the α_{2A}-AR and α_{2B}-AR subtypes were exclusively localized to the cell surface, the α_{2C}-AR was detected on both the lateral subdomain and in an intracellular compartment. The existence of an intracellular pool of the α_{2C}-AR was reported previously in other renal cell lines (von Zastrow *et al.*, 1993). Since the same antibody was utilized to detect all three α_2-AR subtypes, the distinct distribution of the α_{2C}-AR represents the property conferred by the α_{2C}-AR subtype structure, rather than a property of a unique antibody preparation.

In order to identify the intracellular compartments where the α_{2C}-AR is retained, the fluorescent profile of the α_{2C}-AR was compared with calnexin, a marker for the endoplasmic reticulum (Le *et al.*, 1994), and mannosidase II, a marker for the trans-Golgi network (Moremen *et al.*, 1991). Since the α_{2C}-AR overlaps, in part, the immunofluorescence profile of both marker proteins, the interpretation of these studies is either that the intracellular α_{2C}-AR localization spans a number of cellular compartments, including the endoplasmic reticulum and trans-Golgi network or, alternatively, that this receptor subtype exists in a unique intracellular compartment whose immunofluorescent profile partially overlaps that revealed by both calnexin and mannosidase II staining.

Based on the biochemical and morphological studies of receptors evaluated at steady-state, all three subtypes of the α_2-AR are localized to the lateral subdomain of the basolateral surface of MDCK II cells. The next goal was to determine how each α_2-AR subtype achieves its steady-state localization: is it *via* a direct delivery of each receptor subtype to the basolateral surface or *via* random receptor delivery and subsequent selective retention on the basolateral membrane?

In order to examine the delivery of the α_2-AR, newly synthesized proteins were metabolically labeled with [^{35}S]-cysteine/methionine in polarized MDCK II clonal cell lines and the fraction of each subtype arriving on either the apical or the basolateral surface was quantified by the biotinylation strategies described previously (Keefer and Limbird, 1993). The metabolically-labeled α_2-AR subtypes were isolated using the anti-epitope tag antibody and adsorbtion onto protein A-agarose. The fraction of the α_2-AR reaching the biotinylated surface was isolated by streptavidin-agarose chromatography of the protein A-agarose eluates. Similar to the direct delivery of the α_{2A}-AR to the basolateral surface reported previously (Keefer and Limbird, 1993; Keefer *et al.*, 1994), the α_{2C}-AR was delivered directly to the basolateral membrane at each time point examined. The α_{2B}-AR subtype, however, displayed a unique targeting pattern. Thus, the α_{2B}-AR was detected equally on the apical and basolateral surfaces at early time points after initiating [^{35}S]-cysteine/methionine pulse-labeling of cells. However, after 60 minutes of metabolic labeling, the α_{2B}-AR began to accumulate on the basolateral surface and to disappear from the apical membrane. This finding, coupled with the basolateral steady-state localization of the α_{2B}-AR, suggested that the α_{2B}-AR is delivered randomly to both cell surfaces, but is preferentially retained on the basolateral membrane. This interpretation was corroborated by demonstration that the half-life of the α_{2B}-AR on the apical surface is only 15–30 minutes, whereas the half-life of the α_{2B}-AR on the basolateral surface ranges from 8–10 hours. Interestingly, comparison of the half-lives of α_{2A}-AR, α_{2B}-AR and α_{2C}-AR subtypes on the basolateral surface revealed that these three subtypes disappear from the basolateral membrane with similar half-lives of about 8–10 hours.

In summary, our findings indicate that each of the three subtypes of α_2-AR has a different trafficking itinerary. The α_{2A}-AR is delivered directly to the basolateral membrane, where it is almost exclusively present at steady-state. The α_{2C}-AR subtype also is directly delivered to the basolateral surface; however, at steady-state, this receptor subtype is distributed between the basolateral cell surface and an intracellular compartment(s). Unlike the α_{2A}-AR and α_{2C}-AR, the α_{2B}-AR is inserted randomly into both the apical and basolateral surface in MDCK II cells, but is preferentially retained on the basolateral membrane, thus leading to its steady-state localization there. The potential functional consequences of such differing targeting itineraries for each α_2-AR subtype in renal epithelial, or any other polarized cells, are yet to be determined.

References

Graeser, D. and Neubig, R.R. (1993). *Mol. Pharmacol.*, **43**, 434–443.
Guyer, C.A., Horstman, D.A., Wilson, A.L., Clark, J.D., Cragoe, E.J., Jr. and Limbird, L.E. (1990). *J. Biol. Chem.*, **265**, 17307–17317.
Keefer, J.R. and Limbird, L.E. (1993). *J. Biol. Chem.*, **268**, 11340–11347.
Keefer, J.R., Kennedy, M.E. and Limbird, L.E. (1994). *J. Biol. Chem.*, **269**, 16425–16433.
Kleuss, C., Hescheler, J., Ewel, C., Rosenthal, W., Schultz, G. and Wittig, B. (1991). *Nature*, **353**, 43–48.
Kleuss, C., Scherubl, H., Hescheler, J., Schultz, G. and Wittig, B. (1992). *Nature*, **358**, 424–426.
Kleuss, C., Scherubl, H., Hescheler, J., Schultz, G. and Wittig, B. (1993). *Science*, **259**, 832–834.
Kobilka, B.K., Matsui, H., Kobilka, T.S., Yang-Feng, T.L., Francke, U., Caron, M.G., Lefkowitz, R.J. and Regan, J.W. (1987). *Science*, **238**, 650–656.
Lanier, S.M., Hess, H.J., Grodski, A. and Graham, R.M. (1986). *Mol. Pharmacol.*, **29**, 219–227.
Lanier, S.M., Downing, S., Duzic, E. and Homcy, C.J. (1991). *J. Biol. Chem.*, **266**, 10470–10478.
Le, A., Steiner, J.L., Ferrel, G.A., Shaker, J.C. and Sifers, R.N. (1994). *J. Biol. Chem.*, **269**, 7514–7519.
Limbird, L.E. (1989). *FASEB J.*, **2**, 2686–2695.
Moremen, K.W., Touster, O. and Robbins, P.W. (1991). *J. Biol. Chem.*, **266**, 16876–16885.
Muntz, K., Sternweis, P.C., Gilman, A.G. and Mumby, S.M. (1992). *Mol. Biol. Cell*, **3**, 49–61.
Sung, C.H., Davenport, C.M. and Nathans, J. (1993). *J. Biol. Chem.*, **268**, 26645–26649.
von Zastrow, M., Link, R., Daunt, D., Barsh, G. and Kobilka, B (1993). *J. Biol. Chem.*, **268**, 763–766.
Welsh, M.J., Anderson, M.P., Rich, D.P., Berger, H.A., Denning, G.M., Ostedgaard, L.S., Sheppard, D.N., Cheng, S.H., Gregory, R.J. and Smith, A.E. (1992). *Neuron*, **8**, 821–829.
Wozniak, M. and Limbird, L.E. (1996). *J. Biol. Chem.*, **271**, 5017–5024.
Zeng, D., Harrison, J.K., D'Angelo, D.D., Barber, C.M., Tucker, A.L., Lu, Z. and Lynch, K.R. (1990). *Proc. Natl. Acad. Sci. USA*, **87**, 3102–3105.

INTRACELLULAR α_{2A}-ADRENERGIC RECEPTORS IN NEURONS AND GT1 NEUROSECRETORY CELLS

AMY LEE and KEVIN R. LYNCH

Neuroscience Graduate Program and Department of Pharmacology, University of Virginia Health Sciences Center, Charlottesville, VA 22908, USA

We described previously the localization of the A-subtype of α_2-adrenergic receptor (α_{2A}-AR) to large, vesicle-like compartments in neurons of rat brain and the immortalized hypothalamic cell line, GT1. In this study, we sought to characterize this α_{2A}-AR-enriched compartment using electron microscopy and double-label immunofluorescence. In GT1 cells, α_{2A}-AR-immunoreactive structures were larger and extended further into neuritic processes than in primary cultures of hippocampal neurons in which intracellular α_{2A}-AR-immunoreactivity was restricted to cell bodies and proximal dendrites. The α_{2A}-AR-enriched compartment did not colocalize with transferrin receptor- or lgp120-immunofluorescence, indicating that α_{2A}-AR-immunoreactivity is not concentrated in early endosomes or lysosomes, respectively. Furthermore, the subcellular distribution of α_{2A}-AR immunofluorescence was distinct from that of mannose 6-phosphate receptor, a marker of late endosomes. The inability of an α_2-AR agonist to alter the pattern of α_{2A}-AR immunofluorescence suggests that the α_{2A}-AR-immunoreactive structures are not analogous to endosomes involved in agonist-promoted internalization. In GT1 cells, α_{2A}-AR-immunoreactive-structures colocalize with the motor protein, kinesin, and appear to be associated with portions of endoplasmic reticulum that are also α_{2A}-AR-immunoreactive at the ultrastructural level. We conclude that intracellular α_{2A}-ARs accumulate in a non-lysosomal compartment that may be involved in receptor trafficking and/or endocytic functions distinct from early or late endosomes.

KEY WORDS: α_2-adrenoceptor, receptor localization, neurons

INTRODUCTION

Although neurotransmitters and other neuroactive substances are thought to elicit their biological effects primarily through receptors in the plasma membrane, many receptors have been localized to intracellular compartments in neurons. Immunohistochemical studies have shown that δ-opioid receptor-immunoreactivity is compartmentalized in punctate structures within the cytoplasm of cell bodies and dendrites of rat brainstem neurons (Arvidsson *et al.*, 1995) as well as in large dense-core vesicles in axon terminals of spinal cord neurons (Cheng *et al.*, 1996). Using antibodies that recognize specifically the A-subtype of α_2-adrenergic receptor (α_{2A}-AR), our laboratory has found α_{2A}-AR-immunoreactivity to accumulate in large spherical structures, which we have shown to be intracellular by confocal microscopy, in neurons in rat brain (Rosin *et al.*, 1993;

Correspondence to: Kevin R. Lynch, Dept. of Pharamcology, Box 448, University of Virginia Health Sciences Center, 1300 Jefferson Park Avenue, Charlottesville, VA 22908, USA. Fax# (804) 982-3878, Phone# (804) 924-2840.

Guyenet *et al.*, 1994). Electron micrographs revealed that much of the intracellular α_{2A}-AR-immunoreactivity was associated with vesicular structures in cell soma and in axon terminals (Rosin *et al.*, 1995, Talley *et al.*, 1996).

Such intracellular compartments enriched with neurotransmitter receptors may be involved in a number of functions including biosynthesis, subcellular transport, or endocytosis. Although much has been learned about membrane trafficking in neurons, little is known about the neuronal cell biology of neurotransmitter receptors in particular. Thus, characterizing these intracellular, receptor-enriched structures may provide insight into fundamental mechanisms underlying the processing and/or transport of neurotransmitter receptors in neuronal cell systems.

In this study, we employed immunofluorescence and electron microscopy to determine if the α_{2A}-AR-immunoreactive structures exhibited characteristics of intracellular compartments defined previously. We used a neuronal cell line, GT1, that we have shown to express endogenous α_{2A}-ARs by radioligand binding and northern blot analyses (Lee *et al.*, 1995). Additionally, we studied the characteristics of α_{2A}-AR-immunoreactivity in hippocampal neurons in culture since these cells have well-defined dendrites and axons and thus more closely represent the terminally differentiated state of neurons *in vivo*.

MATERIALS AND METHODS

Cell culture

GT1 cells were grown in DMEM/F12 containing 5% dialyzed fetal bovine serum, 5% dialyzed horse serum, and 100 µg/ml penicillin/streptomycin. For immunofluorescence, cells were plated on glass coverslips coated with Matrigel (Collaborative Research) and incubated in serum-free medium overnight. Hippocampal neurons were provided generously by the laboratory of Dr. Gary Banker (University of Virginia) and were cultured on poly-lysine coated glass coverslips as described previously (Goslin and Banker, 1991).

Immunofluorescence

Coverslips of GT1 cells or hippocampal neurons were fixed in 4% paraformaldehyde/ 0.1 M phosphate buffer, permeabilized with 0.1% Triton X-100, and blocked with 3% normal goat serum in phosphate buffered saline (PBS). Cells were incubated overnight at 4°C in a rabbit polyclonal α_{2A}-AR-specific antibody (5 µg/ml, preadsorbed for 2 h with a 10-fold excess of glutathione-S-transferase (GST)) and subsequently in either Texas Red- or FITC-conjugated goat anti-rabbit antibody (3.25 µg/ml). For double label-immunofluorescence, cells were incubated for 1 h at room temperature in primary antibodies diluted in 1% goat serum/PBS, rinsed three times, and incubated for 1 h in either Texas Red- or FITC-conjugated goat anti-mouse antibodies prior to processing with the α_{2A}-AR antibody.

Immunoelectron microscopy

GT1 cells were fixed for 1 h at 37°C in a periodate/lysine/ 2% paraformaldehyde solution, permeabilized for 30 min with 0.02% saponin/2% BSA/PBS (buffer A), and incubated with α_{2A}-AR antibody (2.5 µg/ml, preadsorbed for 2 h with a 10-fold excess of GST) at 4°C overnight. Cells were rinsed in buffer A and incubated in biotinylated goat anti-rabbit antibodies (7.5 µg/ml) for 1 h at room temperature. Coverslips were rinsed in buffer A and post-fixed with 2.5% glutaraldehyde/5% sucrose/0.1 M phosphate buffer. Cells were rinsed in 7.5% sucrose/0.1 M phosphate buffer and subsequently in 7.5% sucrose/50 mM Tris-HCL, pH 7.4. Immunoperoxidase reaction product was detected with the Vectastain ABC kit (Vector Laboratories, Burlingame, CA). Processing for electron microscopy was achieved by incubating coverslips in 1% osmium tetroxide/0.1 M phosphate buffer, pH 7.4 for 30 min at room temperature, staining *en bloc* with 0.5% uranyl acetate for 1 h in the dark, and dehydrated in a series of ethanol steps. Coverslips were flat-embedded in Epon, removed, and cells to be sectioned (50 nm) were selected by light microscopy.

Antibodies

The characterization of the rabbit polyclonal α_{2A}-AR antibody was described previously (Rosin *et al.*, 1993). The authors are grateful for the gifts of: human transferrin receptor monoclonal antibodies (from Dr. Ian Trowbridge, Salk Institute), rabbit polyclonal mannose 6-phosphate receptor antibodies (from Dr. Bill Brown, Cornell University), kinesin H1 monoclonal antibodies (from Dr. Kevin Pfister, University of Virginia) and monoclonal lgp120 antibodies (from Dr. Sam Green, University of Virginia). Fluorescent and biotinylated secondary antibodies (affinity-purified) were obtained from Vector Laboratories.

RESULTS

α_{2A}-AR-immunoreactivity in GT1 cells and hippocampal neurons

α_{2A}-AR-immunoreactivity exhibits a similar subcellular pattern in GT1 cells and hippocampal neurons in culture (Figure 1). In both, α_{2A}-AR-immunoreactivity is punctate and distributed throughout extranuclear regions of the cell body and extends into neuritic processes of GT1 cells and proximal dendrites but not axons of hippocampal neurons. However, immunoreactive puncta are generally smaller and more numerous in hippocampal neurons than in GT1 cells. These data are consistent with the heterogeneity of α_{2A}-AR-immunoreactive structures that we have observed in neurons of rat brain (Talley *et al.*, 1996).

α_{2A}-AR-immunoreactive structures are not early endosomes

To test the hypothesis that the intracellular punctate structures represent endocytosed receptors, we examined whether α_{2A}-AR-immunoreactivity colocalized with markers of early and late endocytic compartments. For example, transferrin receptor is a protein that accumulates in early endosomes in a number of cell systems. To determine if α_{2A}-AR-immunoreactivity is associated with early endosomes, GT1 cells were double-labeled with antibodies against the α_{2A}-AR and transferrin receptor. While both transferrin receptor- and α_{2A}-AR-immunoreactivities were distinctly punctate and restricted primarily to extranuclear regions of the cell soma, α_{2A}-AR-immunoreactivity did not overlap with that

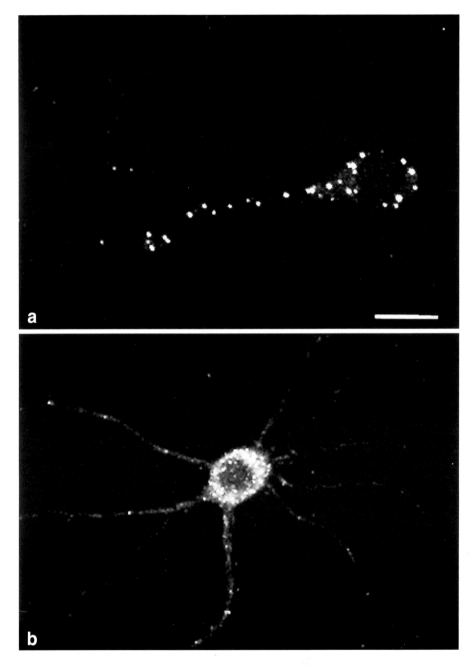

FIGURE 1 Intracellular α_{2A}-AR-immunoreactivity in GT1 cells and rat hippocampal neurons in culture. GT1 cells (a) and hippocampal neurons (b) were immunolabeled with the α_{2A}-AR antibody as described in 'Materials and Methods'. Images represent composites of 9–11 optical sections taken at 0.5–1 μm intervals along the z-axis using a confocal microscope. Scale bar, 7 μm.

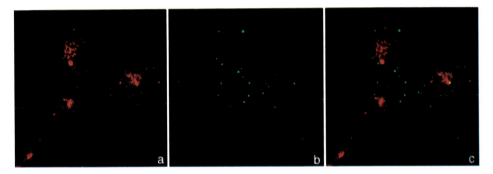

FIGURE 2 α_{2A}-AR-immunoreactive structures do not colocalize with a marker of early endosomes. GT1 cells were double-labeled with α_{2A}-AR- and transferrin receptor antibodies. Texas Red- and FITC-conjugated secondary antibodies were used to achieve transferrin receptor- (a) and α_{2A}-AR- (b) immunofluorescence, respectively. Shown are confocal images representing a single optical section through a group of cells. Note that transferrin receptor-immunoreactivity does not overlap with that for the α_{2A}-AR (c).

for transferrin receptor (Figure 2). In addition, since α_{2A}-AR-immunoreactive structures did not colocalize with early endocytic compartments labeled by fluid phase tracers (not shown), we conclude that the intracellular α_{2A}-AR-immunoreactivity does not accumulate in early endosomes.

Differences between α_{2A}-AR-immunoreactive structures and late endosomes

Late endosomes in many cell types are characterized by the presence of mannose 6-phosphate receptors. In GT1 cells, mannose 6-phosphate receptor-immunoreactivity is clustered at the microtubule organizing center (Figure 3a). This distribution is critically dependent on the presence of intact microtubules since treatment with nocodazole, an agent that disrupts microtubules, caused mannose 6-phosphate receptor-immunoreactive structures to disperse throughout the cell (Figure 3b). By contrast, α_{2A}-AR-immunoreactive structures are not clustered in any region of the cell (Figure 3c) and are not affected by nocodazole (Figure 3d). This difference in subcellular distribution and microtubule dependence suggests that α_{2A}-AR-immunoreactive structures are distinct from late endosomes.

α_{2A}-AR-immunoreactivity is not associated with lysosomes

To determine if α_{2A}-AR-immunoreactivity accumulates in lysosomes, cultured rat hippocampal neurons were double-labeled with antibodies directed against rat lysosomal glycoprotein 120 (lgp 120) and the α_{2A}-AR. lgp120-immunofluorescence was punctate and distributed in somatodendritic regions of the cell, as was α_{2A}-AR-immunofluorescence, but overlay of the images indicated little colocalization of the two proteins (Figure 4). We have found also that α_{2A}-AR-immunoreactivity exhibits a distinct subcellular distribution from that of lysosomal markers in GT1 cells (Lee and Lynch, unpublished). These results suggest that the α_{2A}-AR-immunoreactive structures are not lysosomes.

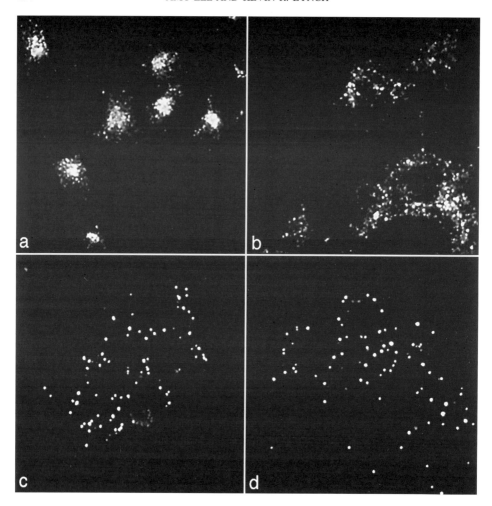

FIGURE 3 Late endosomes exhibit a subcellular localization dependent on intact microtubules distinct from that of α_{2A}-AR-immunoreactive compartments. GT1 cells were immunolabeled with mannose 6-phosphate receptor antibodies to mark late endosomes (a, b) or α_{2A}-AR antibodies (c, d). The clustering of mannose 6-phosphate receptor at the microtubule organizing center (a) is disrupted by nocodazole treatment (10 µM for 1 hour) (b). By contrast, α_{2A}-AR-immunoreactive structures are not clustered in any region of the cell (c) and are not affected by nocodazole (d).

Distribution of α_{2A}-AR-immunoreactive structures is not affected by an α_2-AR agonist

Many receptors such as the β-adrenergic receptor undergo agonist-promoted internalization. Activation of these receptors by agonist leads to rapid removal from the cell surface and subsequent accumulation in early endosomes (von Zastrow and Kobilka, 1992). If the α_{2A}-AR-immunoreactive structures were analogous to these compartments, then treatment of GT1 cells with the specific α_2-AR agonist, UK14304, should increase the number of intracellular structures labeled by the α_{2A}-AR antibody. Since α_{2A}-AR-immunoreactive structures are observed in the absence of agonist and since short (5 min) and long (30 min–

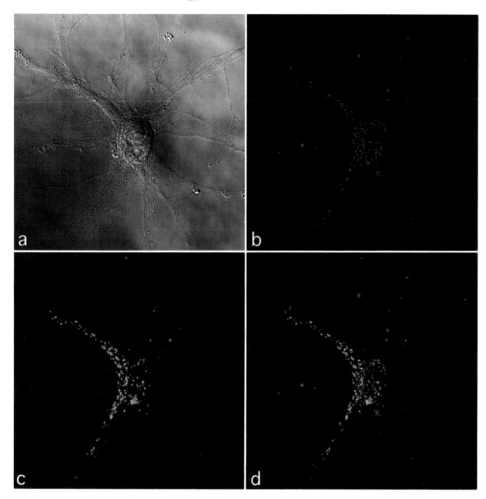

FIGURE 4 α_{2A}-AR-like immunoreactivity is not associated with lysosomes. Cultured rat hippocampal neurons were double-labeled with antibodies directed against rat lysosomal glycoprotein 120 (lgp120) and the α_{2A}-AR. Fluorescein and rhodamine optics were used to visualize lgp120- (c) and α_{2A}-AR- (b) immunofluorescence, respectively. Shown are a differential contrast image of the cell (a) and composites of 12 optical sections taken through the z-axis using a confocal microscope (b, c). Overlay of images in b and c indicates little colocalization of the two proteins (d).

1 hour) incubations with agonist (1 µM) did not alter the number or subcellular distribution of immunolabeled puncta in GT1 cells (not shown), we conclude that the α_{2A}- AR-immunoreactive structures do not contain receptors undergoing agonist-promoted internalization.

Association with immunolabeled portions of endoplasmic reticulum

In previous ultrastructural analyses, α_{2A}-AR-immunoreactivity was found to accumulate in large multivesicular structures in cell bodies and axon terminals of neurons in rat brain

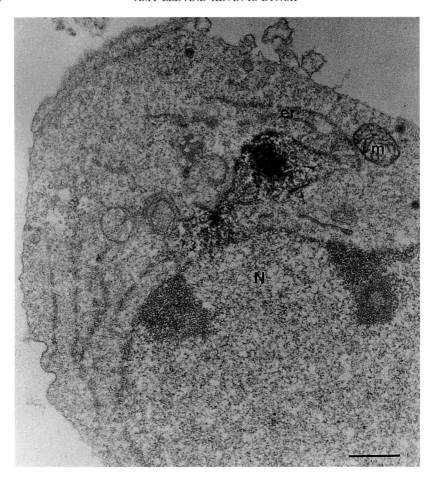

FIGURE 5 α_{2A}-AR-immunoreactive structures are associated with immunolabeled portions of endoplasmic reticulum. GT1 cells were immunolabeled with α_{2A}-AR antibodies and processed for electron microscopy as described in 'Materials and Methods'. α_{2A}-AR-immunoreactivity was found consistently on large structures (arrowhead) as well as closely apposing regions of endoplasmic reticulum (arrow). Abbreviations: N, nucleus; m, mitochondria; er, endoplasmic reticulum. Scale bar, 1 µm.

(Rosin et al., 1995, Talley et al., 1996). In addition, α_{2A}-AR-immunoreactivity was associated also with large, possibly vesicular structures closely apposed to regions of labeled endoplasmic reticulum (Milner, Rosin and Lynch, unpublished). Although we did not observe multivesicular structures labeled with the α_{2A}-AR antibody in GT1 cells, α_{2A}-AR-immunoreactivity was found consistently on large structures near the endoplasmic reticulum (Figure 5), although diffusion of the immunoperoxidase reaction product tended to obscure the morphology of α_{2A}-AR-immunoreactive structures. This pattern of immunoreactivity was not seen when the antibody was preadsorbed with the immunogenic receptor-fusion protein (not shown). While we can only speculate as to the functional significance of these structures, they do not appear to be similar to endocytic compartments described previously in neurons (Parton et al., 1992).

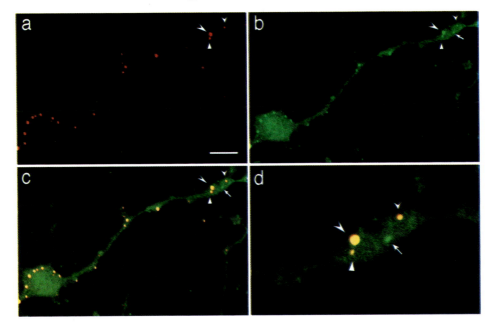

FIGURE 6 α_{2A}-AR-like immunoreactivity colocalizes with the motor protein kinesin. GT1 cells were double-labeled with antibodies against kinesin and α_{2A}-AR. Excitation and emission wavelengths were set for rhodamine for α_{2A}-AR immunofluorescence (a) and fluorescein for kinesin (b). Images represent composites of 11 optical sections taken at 1 μm intervals along the z-axis using a confocal microscope. Superimposed image reveals strong colocalization of kinesin and α_{2A}-AR-immunoreactivity represented in yellow (c). (d) Magnified view of process shown in a–c. Scale bar, 10 μM (a–c), 3 μM (d).

Colocalization with the motor protein kinesin

Although we have not been successful thus far in identifying proteins associated with the α_{2A}-AR-immunoreactive structures that would be indicative of function, we have found a one-to-one correspondence of α_{2A}-AR-immunoreactivity and that of the motor protein, kinesin. In GT1 cells that were double-labeled immunofluorescently with antibodies against kinesin and α_{2A}-AR, we found that both kinesin- and α_{2A}-AR-immunoreactivity accumulated in the same structures (Figure 6). While all α_{2A}-AR-immunoreactive structures were double-labeled with the kinesin antibody, many kinesin-immunoreactive structures did not exhibit α_{2A}-AR-immunoreactivity, which is consistent with the association of kinesin with a variety of membrane-bound organelles. While it is premature to draw conclusions regarding the functional significance of these data, the association of α_{2A}-AR-immunoreactive structures with a motor protein such as kinesin could underlie a mechanism by which α_{2A}-ARs are transported to particular subcellular domains.

DISCUSSION

The development of antibodies that recognize specific neurotransmitter receptors in immunohistochemical preparations has allowed detailed analyses of the distribution of these signaling molecules at the neuroanatomic and subcellular levels. An interesting feature of a number of receptors is their accumulation in intracellular, vesicular structures in neurons. While mapping immunohistochemically the A and C-subtypes of α_2-ARs in the rat central nervous system, our laboratory has found that both subtypes, particularly the A-subtype, are associated with such intracellular compartments (Rosin et al., 1993; Rosin et al., Talley et al., 1996). Although much research has focused on the actions of receptors situated in the plasma membrane, little is known about how neurotransmitters are transported to and actively maintained at their sites of action. Therefore, we sought to characterize the α_{2A}-AR-immunoreactive compartments in relation to other membrane-bound organelles.

Due to the technical difficulties of characterizing intraneuronal α_{2A}-ARs in rat brain, we used a neuronal cell line, GT1, and cultured rat hippocampal neurons in this study. Although the pattern of intracellular α_{2A}-AR-immunoreactivity was similar in both GT1 cells and hippocampal neurons, α_{2A}-AR-immunoreactive structures differed in size and number between the two cell cultures. These data provide an interesting parallel to what we have observed in rat brain, where the size and number of α_{2A}-AR-immunoreactive structures within a neuron was found to vary according to neuroanatomic loci. For example, neurons in the locus coeruleus contain numerous, relatively small immunoreactive puncta whereas these intracellular structures are very large and few in number in some hypothalamic neurosecretory neurons (Talley et al., 1996). At present it is unclear as to why the α_{2A}-AR-immunoreactive structures should be so heterogeneous among neuronal cell types, although one explanation may lie in the inherent differences in subcellular handling of α_{2A}-ARs that may exist between neurons.

Regarding the identity of the α_{2A}-AR-immunoreactive structures, we first hypothesized that these compartments may represent some intermediate in the endocytic pathway, since neurotransmitter receptors, like all plasma membrane proteins, undergo some form of endocytosis and/or recycling in their lifetime. A major prediction of this hypothesis is that α_{2A}-AR-immunoreactivity should colocalize with markers of known endocytic compartments. However, we found that the α_{2A}-AR-immunoreactive structures were not immunolabeled with antibodies against early endosomal and lysosomal proteins. The inability of α_2-AR agonists to alter the distribution of intracellular α_{2A}-AR-immunoreactivity suggests also that the compartment is probably not involved in agonist-promoted internalization of receptors to early endocytic compartments. Furthermore, the α_{2A}-AR-immunoreactive compartment could be distinguished from late endosomes based on the microtubule-independence of its subcellular localization. We have also preliminary evidence indicating that the α_{2A}-AR-immunoreactive structures are unlikely to be carrier vesicles, structures that are thought to carry endocytosed material from late endosomes to lysosomes (Lee and Lynch, unpublished). From these data, we conclude that intracellular α_{2A}-AR-immunoreactivity is not associated with any compartment described previously in the endocytic pathway, although we cannot eliminate the possibility that these structures may represent a novel intermediate involved in endocytic processing.

Although immunogold labeling will be necessary to resolve the morphology of the α_{2A}-AR-immunoreactive structures at the ultrastructural level, our preliminary results suggest that the intracellular compartments labeled with the α_{2A}-AR antibody are associated closely with the endoplasmic reticulum (ER). Given the role of the ER as a site of biosynthetic processing, it is interesting to speculate that the α_{2A}-AR-immunoreactive structures might somehow be involved in the packaging of newly synthesized receptors for subcellular transport. The colocalization of α_{2A}-AR-immunoreactivity with that of kinesin, a motor protein believed to be involved in the anterograde transport of membrane-bound organelles, supports this possibility also. While the immunofluorescent localization of motor proteins is not always a reliable indicator of function (Lippincott-Schwartz et al., 1995), we have preliminary evidence that the α_{2A}-AR-immunoreactive structures are not associated with the putative retrograde motor, dynein (Lee and Lynch, unpublished). Given the prominent role of α_{2A}-ARs as presynaptic receptors in the central nervous system, an intriguing possibility is that the α_{2A}-AR-immunoreactive structures extending into the neuritic processes of GT1 cells (Figure 6) may be involved in the anterograde transport of receptors to presynaptic sites of action.

In summary, we described the properties of intracellular α_{2A}-ARs in cultured neurons and a neuronal cell line using a subtype-selective antibody. α_{2A}-AR-immunoreactivity is associated with intracellular compartments that differ from early and late endosomes and lysosomes. The α_{2A}-AR-immunoreactive structures are not similar to early endocytic compartments involved in agonist-promoted internalization since they are observed in the absence of agonist stimulation. While further characterization of these compartments is required, α_{2A}-AR-immunoreactive structures may be involved in endocytic functions distinct from those subserved by lysosomes and early and late endosomes or alternatively, in the biosynthetic processing and/or transport of the receptor to particular subcellular domains.

ACKNOWLEDGEMENTS

We acknowledge the contributions of reagents and helpful advice from our colleagues at the University of Virginia: Drs. David Castle, Kevin Pfister, Sam Green, and Gary Banker. In addition, we thank Paula Falk from Dr. Oswald Steward's laboratory for assistance with electron microscopy. This work was supported by grants from the National Institutes of Health (T32 HL07284 to A.L. and R01 DA07216 to K.R.L.) and the American Heart Association.

References

Arvidsson, U., Dado, R.J., Riedl, M., Lee, J., Law, P.Y., Loh, H.H., Elde, R. and Wessendorf, M,W, (1995). δ-Opioid receptor immunoreactivity: distribution in brainstem and spinal cord, and relationship to biogenic amines and enkephalin. *J. Neurosci.*, **15**, 1215–1235.

Cheng, P.Y., Svingos, A.L., Wang, H., Clarke, C.L., Jenab, S., Beczkowska, I.W., Inturrisi, C.E. and Pickel, V.M. (1996). Ultrastructural immunolabeling shows prominent presynaptic vesicular localization of δ-opioid receptor within both enkephalin- and nonenkephalin-containing axon terminals in the superficial layers of the rat cervical spinal cord. *J. Neurosci.*, **15**, 5976–5988.

Goslin, K. and Banker, G. (1991). Rat hippocampal neurons in low-density culture. *Culturing Nerve Cells*, pp. 251–281. Cambridge, MA: MIT Press.

Guyenet, P.G., Stornetta, R.L., Riley, T., Norton, F.R., Rosin, D.L. and Lynch, K.R. (1994). α2-Adrenergic receptors are present in lower brainstem catecholaminergic and serotonergic neurons innervating sympathetic preganglionic cells. *Brain Res.*, **638**, 285–294.

Lee, A., Talley, E.M., Rosin, D.L. and Lynch, K.R. (1995). Characterization of α_{2A} adrenergic receptors in GT1 neurosecretory cells. *Neuroendocrinology*, **62**, 215–225.

Lippincott-Schwartz, J., Cole, N.B., Marotta, A., Conrad, P.A. and Bloom, G.S. (1995). Kinesin is the motor for microtubule-mediated Golgi-to-ER membrane traffic. *J. Cell Biol.*, **128**, 293–306.

Parton, R.G., Simons, K. and Dotti, C.G. (1992). Axonal and dendritic endocytic pathways in cultured neurons. *J. Cell Biol.*, **119**, 123–137.

Rosin, D.L., Zeng, D., Stornetta, R.L., Riley, T., Okusa, M.D., Guyenet, P.G. and Lynch, K.R. (1993). Immunohistochemical localization of α_{2A}-adrenergic receptors in catecholaminergic and other brainstem neurons in the rat. *Neuroscience*, **56**, 139–155.

Rosin, D.L., Talley, E.M., Milner, T.A., Guyenet, P.G. and Lynch, K.R. (1995). Immunohistochemical localization of α_{2A}-adrenergic receptors in rat forebrain: light (LM) and electron microscopy (EM). *Soc. Neurosci. Abstr.*, **21(2)**, 1622.

Rosin, D.L., Talley, E.M., Lee, A., Stornetta, R.L., Gaylinn, B., Guyenet, P.G. and Lynch, K.R. (1996). Distribution of α_{2C}-adrenergic receptor-like immunoreactivity in the rat central nervous system. *J. Comp. Neurol.*, in press.

Talley, E.M., Rosin, D.L., Lee, A., Guyenet, P.G. and Lynch, K.R. (1996). Distribution of α_{2A}-adrenergic receptor-like immunoreactivity in the rat central nervous system. *J. Comp. Neurol.*, in press.

von Zastrow, M. and Kobilka, B.K. (1992). Ligand-regulated internalization and recycling of human β_2-adrenergic receptors between the plasma membrane and endosomes containing transferrin receptors. *J. Biol. Chem.*, **267**, 3530–3538.

IMMUNOLOCALIZATION OF NATIVE α_2-ADRENERGIC RECEPTOR SUBTYPES: DIFFERENTIAL TISSUE AND SUBCELLULAR LOCALIZATION

JEREMY G. RICHMAN, YI HUANG and JOHN W. REGAN

Departments of Pharmacology & Toxicology and Physiology, College of Pharmacy, University of Arizona, Tucson, AZ 85721, USA

A number of techniques are available for the study of α_2-adrenergic receptor (α_2AR) cellular and subcellular localization. Although each of these techniques are limited to some capacity, they have provided valuable information with regard to subtype specific differences in both basal (unstimulated) conditions as well as differences in the response of the receptors following exposure to agonist. We have studied the potential for differential α_2-AR cellular and subcellular localization in a number of *in vitro* and *in vivo* settings. Our findings indicate that α_2-AR immunoreactivity in transiently transfected COS-7 cells, differentially localizes under basal conditions (i.e. the α_{2A}- and α_{2B}-ARs are primarily on the plasma membrane while the α_{2C}-AR appears to maintain a position within the periplasmic space). Following stimulation by agonist the α_{2A}- and α_{2B}-ARs appear to undergo sequestration of immunoreactivity while the α_{2C}-AR does not change. In addition, *in vivo* studies of the human ocular ciliary body, rat spinal cord and rat aortic smooth muscle revealed that the α_2-ARs are differentially expressed, may be co-expressed within the same cells (in the case of rat spinal cord neurons) and may undergo a time dependent decrease in immunoreactivity (in the case of rat aortic smooth muscle cells).

Three α_2-adrenergic receptors (AR) have been identified to date based on the pharmacological profile of ligand preference, and have been named α_{2A}, α_{2B} and α_{2C} (Bylund *et al.*, 1994). These receptor subtypes have been further classified on a molecular basis, as a function of their sequence homology and chromosomal localization. Previous studies with human cells have referred to these receptor subtypes as α_2-C10, α_2-C2, and α_2-C4, corresponding to the chromosome on which their gene is located (Kobilka *et al.*, 1987; Lomasney *et al.*, 1990; Regan *et al.*, 1988). Because the three α_2-AR subtypes produce their actions in response to the binding of the endogenous catecholamines norepinephrine and epinephrine, a number of hypotheses as to the physiological relevance of having three α_2-adrenergic receptor subtypes have been proposed and there are currently a number of research paradigms designed to elucidate their roles. Relevant, is the fact that the receptors differ in both their primary amino acid sequence and corresponding gene encoding sequence, lending to the potential to be differentially expressed in tissues (Kobilka *et al.*, 1987; Lomasney *et al.*, 1990; Regan *et al.*, 1988; Renouard *et al.*, 1994). Also relevant are the subtypes' respective affinity and selectivity for both agonists and antagonists (Bylund *et al.*, 1994; Regan and Cotecchia, 1992; Renouard *et al.*, 1994), their ability

---

Correspondence to: John W. Regan, Depts. of Pharmacology & Toxicology and Physiology, College of Pharmacy, University of Arizona, Tucson, AZ 85721, USA. e-mail: regan@tonic.pharm.arizona.edu. Tel: 520-626-2181, Fax: 520-626-4063

to couple differentially to various second messengers (Pepperl and Regan, 1993; Chabre et al., 1994; Eason and Liggett, 1993), and differential sensitivities to phosphorylation and desensitization in response to agonist occupancy (Kurose and Lefkowitz, 1994). An additional parameter wherein the subtypes appear to show subtype selectivity concerns their subcellular localization and trafficking. For example, Von Zastrow et al. (1992) have shown that the α_2-AR subtypes maintain diverse subcellular distributions in both a steady-state (untreated), as well as an agonist-activated cell. Furthermore, these differences in receptor-subtype cellular trafficking are a function of the tissue in which they are expressed (Huang, 1995; Richman, present findings).

Until recently, the ability to study receptor localization and trafficking has been limited by methodological considerations. For example, initial receptor down-regulation studies of α_2-adrenergic receptors, utilized a technique of whole-cell binding to elucidate potential differences in receptor subcellular localization (Eason and Liggett, 1992). It was found that while the human α_{2A} and α_{2B}-ARs appear to undergo agonist-promoted sequestration and down-regulation, the α_{2C}-AR does not. Furthermore, results seem to indicate that under steady-state conditions, untreated cells stably transfected with the α_{2A} and α_{2C} exhibit a low level of intracellular expression (5.7 ± 0.7 or $0.6 \pm 0.6\%$ sequestration respectively), which is higher for the α_{2B} receptor ($18.8 \pm 2.4\%$ sequestration). As a technique for monitoring receptor trafficking and subcellular localization, however, whole-cell binding is limited by a number of assumptions including the permeability of the ligands involved and the availability of subtype-selective ligands. Immunohistochemistry is an alternative technique that has the potential advantage of being highly selective for individual receptor subtypes.

Recently, immunocytochemical localization has been utilized to study potential subtype differences in the α_2-ARs (Fonseca et al., 1995; Von Zastrow and Kobilka, 1992; Von Zastrow and Kobilka, 1994; Von Zastrow et al., 1992). In contrast to the whole-cell binding results, using confocal immunofluorescence microscopy and antibodies specific to epitope-tagged α_2-ARs, Von Zastrow et al. (1992) showed, that in steady-state (unstimulated) COS-7 and HEK293 cells, the mouse α_{2A}-AR is found primarily on the plasma membrane, while the α_{2C}-AR is, to a large extent, found intracellularly in vesicles. Interestingly, in contrast to the β_2-ARs which undergo agonist promoted internalization, the α_{2A}-AR did not internalize following exposure to agonist (100 μM norepinephrine for 10 min).

To further elucidate potential differences among the α_2-AR subtypes with respect to both their subcellular distribution and response to agonist, we have developed polyclonal antibodies specific to the third intracellular loop of the individual receptor subtypes (Vanscheeuwijck et al., 1993; Huang et al., 1995). It was our hypothesis that differences may exist not only in subtype-specific subcellular distribution, and response to agonist, but any divergence in receptor distribution may also be a function of the specific tissue or transfected cell line in which the receptor(s) are expressed. To study endogenous receptors we chose to investigate cells and tissues in which α_2-adrenergic responses are known to mediate physiological effects, including ciliary body of the eye, spinal cord neurons and aortic smooth muscle cells.

Antibodies were made by inoculating chickens with glutathione-S-transferase fusion proteins which contained portions of the third intracellular loop of each of the α_2AR subtypes, followed by purification of the antibodies from egg yolks (Huang, 1995; Huang

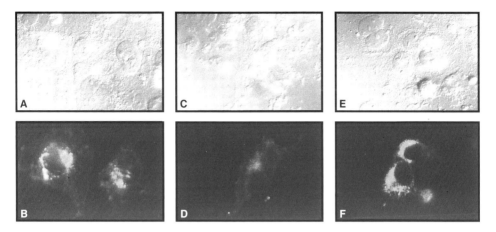

FIGURE 1 Immunofluorescence photomicrograph of COS-7 cells transfected with the DNA encoding human α_{2A} (panels A & B), α_{2B} (panels C & D) and α_{2C} (panels E & F) ARs and labeled with anti-α_{2A} (panel B), anti-α_{2B} (panel D) and anti-α_{2C} (panel F) antibodies. Cells grown on coverslips, fixed in 4% paraformaldehyde and visualized under DIC microscopy (panels A, C & E) and fluorescence microscopy (using and FITC filter; panels B, D and F). Magnification is 850×.

et al., 1995; Vanscheeuwijck *et al.*, 1993). Because little homology exists within the primary sequence of the third intracellular loop of the α_2-ARs, these antibodies are subtype-specific and do not cross-react with other receptors. Furthermore, because they recognize the native receptor, it is possible to do immunohistochemical staining of receptors in primary cultures and intact tissue.

Utilizing these subtype-specific antibodies for immunohistochemical staining, we found that in transiently transfected COS-7 cells, the α_{2A} and α_{2B} immunoreactivity appears to be localized predominantly at the plasma membrane, whereas the α_{2C} immunoreactivity appears to be largely intracellular. Using differential interference contrast (DIC) and fluorescence microscopy it appears that the α_{2C} immunoreactivity is located in punctate, perinuclear vesicles (Figure 1). Furthermore, following the administration of norepinephrine (100 µM) or dexmedetomidine (1 µM), α_{2A} and α_{2B} immunoreactivity, appears to undergo time-dependent internalization, while the α_{2C} does not. Receptor immunoreactivity appears slightly punctate following a 15 min exposure to agonist at 37°C, and following an overnight exposure to agonist (16–18 hours), the receptor immunoreactivity appears to be internalized and concentrated within a perinuclear space, distinct from the cell membrane (as determined using DIC microscopy to visualize the cell membrane; Figures 2a & 2b).

To examine the localization of α_2-AR subtypes within native cells and tissues immunofluorescence microscopy was also performed with rabbit ocular ciliary body, rat spinal cord neurons, and rat and bovine aortic smooth muscle cells. The results indicate that the α_2-ARs are, differentially expressed within distinct tissue types, and further, that the receptors may undergo differential cellular trafficking and cellular localization, as a function of the endogenous tissue.

In cells of the human ocular ciliary body, α_{2B} and α_{2C} immunoreactivity was found, but not α_{2A}, while in the rabbit ocular ciliary body, all three subtypes were immunoreactive

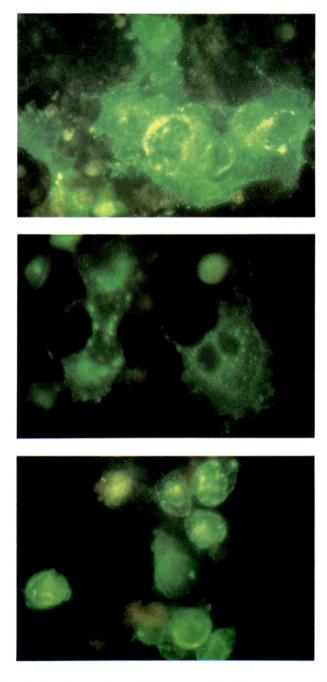

FIGURE 2a Internalization of α_{2A}- and α_{2B}-AR immunoreactivity in transiently transfected COS-7 cells. COS-7 cells transiently transfected with the human α_{2A}-AR and stained with polyclonal antibodies specific for the third intracellular loop of the human α_{2A}-AR (1:1000 dilution) following exposure to agonist (top panel, untreated control; middle panel 15 min dexmedetomidine (1 μM); bottom panel, 16 hour dexmedetomidine (1 mM)).

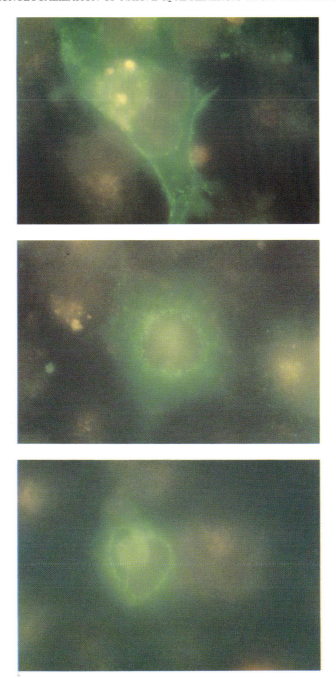

FIGURE 2b Internalization of α_{2A}- and α_{2B}-AR immunoreactivity in transiently transfected COS-7 cells. COS-7 cells transiently transfected with the human α_{2B}-AR and stained with polyclonal antibodies specific for the third intracellular loop of the human α_{2B}-AR (1:50 dilution) following exposure to agonist (top panel, untreated control; middle panel, 15 min dexmedetomidine (1 µM); bottom panel, 16 hour dexmedetomidine (1 µM)).

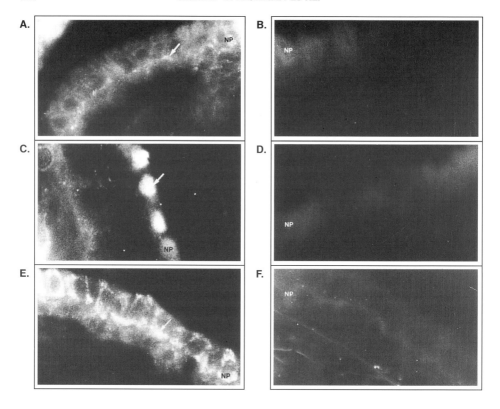

FIGURE 3 Immunofluorescence labeling of the rabbit ciliary process with antibodies raised against the human GST/α_{2A} (panels A & B), GST/α_{2B} (panels C & D) and GST/α_{2C} (panels E & F) fusion proteins. Panels B, D and F show the labeling that was obtained after preincubation of the primary antibodies with the corresponding fusion proteins for 16–18 hours at 4°C. Arrows indicate the labeling in the epithelium. Magnification 990 ×. NP, nonpigmented epithelium. Adapted from Huang et al. (1995).

(Huang, 1995; Huang et al., 1995). Specifically, anti-α_{2B}- and α_{2C}-AR antibodies positively stained the nonpigmented epithelium and ciliary muscle. Comparison of the immunofluoresce microscopy with DIC images suggests that receptors are present in the perinuclear space as well as on the cell surface (Figure 3).

In the rat embryonic spinal cord, α_{2A} and α_{2B} immunoreactivity is present and is colocalized to the same neurons (Huang, 1995). In addition, within these neurons, the α_{2A} immunoreactivity is located primarily on the plasma membrane of the cell body but it is also present in a punctate form in neurites. Similarly, the α_{2B} immunoreactivity is predominantly on the plasma membrane of the cell body, however the staining in the neurites is qualitatively different from that of the α_{2A} and is linear or segmented, rather than punctate (Huang, 1995).

Aortic smooth muscle cells were isolated from bovine (BASM) and rat (RASM) tissue explants from transverse and descending aorta. A monoclonal antibody to smooth muscle α-actin was used to verify that the cells were of a smooth muscle phenotype. Subtype-specific antibodies to the α_2-ARs showed abundant α_{2A} immunoreactivity, whereas, α_{2B}- and α_{2C} immunoreactivity was very low or absent, respectively. With respect to the

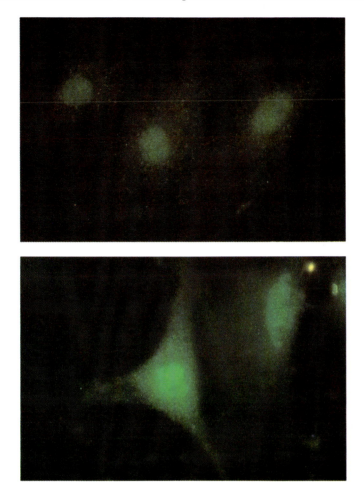

FIGURE 4 Bovine aortic smooth muscle cells (BASM) were fixed in 4% paraformaldehyde, permeabilized using 0.05% saponin and incubated with anti-third intracellular loop α_{2A}-AR antibodies (1:850 dilution). They were then incubated with an FITC labeled secondary rabbit anti-chicken IgG antibody (1:1000 dilution). *Top panel* untreated control. *Bottom panel* cells treated for 16 hours with 1 µM dexmedetomidine.

subcellular localization, the α_{2A} immunoreactivity in the BASM cells is primarily on the plasma membrane. In response to agonist (1 µM dexmedetomidine, 16 hr), α_{2A} immunoreactivity takes on a more punctate appearance and the intensity of the labeling decreases (Figure 4). α_{2B} immunoreactivity did not appear to change following agonist exposure. It is interesting to note however, that contrary to the studies in transfected cells, in aortic smooth muscle the α_{2A} immunoreactivity did not respond at all to short-term agonist treatment. Only following agonist incubations of 2 or more hours did changes take place. This may be indicative of down-regulation. In conclusion, it is apparent that the α_2-AR subtypes display a wide range of heterogeneity with respect to their tissue distribution, and this heterogeneity may be responsible for the diversity of a number of physiological functions mediated by the α_2-ARs.

References

Bylund, D., Eikenberg, D., Hieble, J., Langer, S., Lefkowitz, R.J., Minnemann, K., Molinoff, P., Ruffolo, R. and Trendelenburg, U. (1994). IV. International union of pharmacology nomenclature of adrenoceptors. *Pharmacological Reviews*, **46**, 121–136.

Chabre, O., Conklin, B., Brandon, S., Bourne, H. and Limbird, L. (1994). Coupling of the α_{2A}-adrenergic receptor to multiple G-proteins. *JBC*, **269**, 5730–5734.

Eason, M. and Liggett, S. (1992). Subtype-selective desensitization of α_2-adrenergic receptors. *JBC*, **267**, 25473–25479.

Eason, M. and Liggett, S. (1993). Functional α_2-adrenergic receptor-G_s coupling undergoes agonist-promoted desensitization in a subtype-selective manner. *BBRC*, **193**, 318–323.

Fonseca, M., Button, D. and Brown, D. (1995). Agonist regulation of α_{1B}-adrenergic receptor subcellular distribution and function. *JBC*, **270**, 8902–8909.

Huang, Y. (1995). Location of α_2-adrenergic receptor subtypes using subtype-specific antibodies. University of Arizona. pp. 1–196. PhD. Physiology.

Huang, Y., Gil, D.W., Vanscheeuwijck, P., Stamer, D.W. and Regan, J.W. (1995). Localization of α_2-adrenergic receptor subtypes in the anterior segment of the human eye with selective antibodies. *IOVS*, **36**, 2729–2739.

Kobilka, B.K., Kobilka, T.S., Yang-Feng, T.L., Franke, U., Matsui, H., Caron, M.G., Lefkowitz, R.J. and Regan, J.W. (1987). Cloning, sequencing, and expression of the gene coding for the human platelet α_2-adrenergic receptor. *Science*, **238**, 650–656.

Kurose, H. and Lefkowitz, R.J. (1994). Differential desensitization and phosphorylation of three cloned and transfected α_2-adrenergic receptor subtypes. *JBC*, **269**, 10093–10099.

Lomasney, J., Lorenz, W., Allen, L., King, K., Regan, J.W., Yang-Feng, T.L., Caron, M.G. and Lefkowitz, R.J. (1990). Expansion of the α_2-adrenergic receptor family: cloning and characterization of a human α_2-adrenergic receptor subtype, the gene for which is located on chromosome 2. *PNAS USA*, **87**, 5094–5098.

Pepperl, D. and Regan, J.W. (1993). Selective coupling of α_2-adrenergic receptor subtypes to cyclic AMP-dependent reporter gene expression in transiently transfected JEG-3 cells. *Mol. Pharm.*, **44**, 802–809.

Regan, J.W., Kobilka, T.S., Yang-Feng, T.L., Caron, M.G., Lefkowitz, R.J. and Kobilka, B.K. (1988). Cloning and expression of a human kidney cDNA for an α_2-adrenergic receptor subtype. *PNAS USA*, **85**, 6301–6305.

Regan, J.W. and Cotecchia, S. (1992). The α-adrenergic receptors: new subtypes, pharmacology and coupling mechanisms. In: Brann, M.R. (Ed.) *Molecular Biology of G-Protein-Coupled Receptors*. pp. 76–112. Boston: Birkhauser Boston, Inc.

Renouard, A., Widdowson, P.S. and Millan, M.J. (1994). Multiple α_2-adrenergic receptor subtypes. I. Comparison of [^{3}H]RX821002-labeled rat Rα_{-2A}-adrenergic receptors in cerebral cortex to human Hα_{2A}-adrenergic receptor and other populations of α_2-adrenergic subtypes. *JPET*, **270**, 946–957.

Vanscheeuwijck, P., Huang, Y., Schullery, D. and Regan, J.W. (1993). Antibodies to a human α_2-C10 adrenergic receptor fusion protein confirm the cytoplasmic orientation of the V–VI loop. *BBRC*, **190**, 340–346.

Von Zastrow, M., Daunt, D., Barsh, G. and Kobilka, B.K. (1992). Subtype-specific differences in the intracellular sorting of G protein-coupled receptors. *JBC*, **268**, 763–766.

Von Zastrow, M. and Kobilka, B.K. (1992). Ligand regulated internalization and recycling of human β_2-adrenergic receptors between the plasma membrane and endosomes containing transferrin receptors. *JBC*, **267**, 3530–3538.

Von Zastrow, M. and Kobilka, B.K. (1994). Antagonist-dependent and -independent steps in the mechanism of adrenergic receptor internalization. *JBC*, **269**, 18448–18452.

α_2 ADRENERGIC RECEPTORS IN THE PATHOPHYSIOLOGY AND TREATMENT OF DEPRESSION

HUSSEINI K. MANJI, MARK E. SCHMIDT, FRED GROSSMAN, KARON DAWKINS and WILLIAM Z. POTTER

Departments of Psychiatry and Pharmacology,
Wayne State University School of Medicine,
and Section on Clinical Pharmacology, National Institute of Mental Health

The affective disorders are chronic, severe, recurrent and often life-threatening disorders. The recognition of the significant morbidity and mortality of the severe mood disorders, as well as the growing appreciation that a significant percentage of patients respond poorly to existing treatments has made the task of discovering new therapeutic agents increasingly more important. Several lines of evidence suggest that there are abnormalities of the α_2 adrenergic receptor (α_2AR) system in depression, and that α_2AR antagonists may represent a novel class of antidepressants. In this context, recent studies have shown that idazoxan and mitrazapine are effective antidepressants with remarkably few side effects, even in patients who are resistant to established therapies. These agents hold the potential not only for treating the significant number of patients who do not respond to existing treatments, but also for facilitating in the identification of the biochemical substrates predisposing individuals to severe affective disorders.

KEY WORDS: Affective disorders, depression, norepinephrine, α_2 adrenoceptors, idazoxan, mitrazapine

INTRODUCTION

The severe affective disorders are among our society's most urgent health problems. The disability, chronicity, and social stigma associated with these illnesses exact a very large personal, familial, and economic toll. In recent years, there is a growing appreciation that although treatments do currently exist for the treatment of severe depression, a significant number of patients do not respond at all, while many others are helped but continue to suffer significant morbidity. Additionally, although the standard ADs are generally effective in the treatment of depression, they convey a host of troublesome adverse effects (Rudorfer *et al.*, 1994). Although the search to find more effective, safer and more rapidly acting antidepressants has continued unabated since then, progress in drug development has been very slow, with the exception of the group of serotonin-selective reuptake inhibitors (SSRI's). Although this class of drugs have significantly reduced serious side effects and are thus a useful addition to our therapeutic armamentarium, they represent

---

Correspondence to: Husseini K. Manji, MD, FRCPC, Director, Schizophrenia and Mood Disorders Program, WSU School of Medicine, UHC 9B, 4201 St Antoine Blvd, Detroit, MI 48201, USA. Tel: (313)-966-7544, Fax: (313)-577-5900.

yet another example of somewhat cleaner "me too" drugs, and may even be less effective than the older tricyclics in the treatment of severe depressions (Manji et al., 1995). The recognition of the significant morbidity and mortality of the severe mood disorders, as well as the growing appreciation that a significant percentage of patients respond poorly to existing treatments has made the task of discovering new therapeutic agents that work quickly, potently (Manji et al., 1995), specifically, and with few side effects increasingly more important. One novel class of putative antidepressants are α_2 adrenergic receptor (α_2AR) antagonists; these agents hold the potential not only for treating the significant number of patients who do not respond to existing treatments, but also to facilitate in the identification of the biochemical substrates predisposing individuals to severe affective disorders.

THE NORADRENERGIC SYSTEM IN DEPRESSION

Originally proposed a quarter century ago, the catecholamine hypothesis of affective disorders posited a deficiency of norepinephrine (NE) at critical sites in the CNS in depression (reviewed in Siever, 1987; Manji et al., 1994). These hypotheses were generated from the clinical observation that drugs which deplete CNS monoamines (e.g. reserpine) produced depression in some individuals and from the *in vitro* monoamine reuptake blocking activity of the therapeutically effective tricyclic antidepressants. This postulate has been extensively investigated but has been proven difficult to study experimentally due, at least in part, to the formidable methodological difficulties in assessing CNS noradrenergic function in humans. The sheer volume of research data that have been generated has contributed to apparent controversies in the literature concerning not only the nature of noradrenergic dysfunction but also whether any such dysfunctions exist (Potter and Manji, 1994). We will briefly review, critically appraise, and integrate the research findings of the present status of findings on NE in affective illness, before discussing the role of α_2AR antagonists in the treatment of depression.

Plasma and CSF norepinephrine

Most clinical investigations of plasma NE reveal some degree of elevation which is interpreted as evidence of increased peripheral sympathetic nervous system activity (reviewed in Siever, 1987; Manji et al., 1994). Since plasma NE is determined by a variety of factors, including release, reuptake and degradation, a technique for measuring NE "spillover rates" was designed. At least two groups (reviewed in Siever, 1987; Manji et al., 1994) reported that depressed patients had elevated spillover rates, interpretable as representing the rate of entry of NE to plasma. Studies of the responsiveness of plasma NE to various provocative challenge tests have also been carried out in an effort to provide evidence for dysregulation of the NE system. Thus, in an orthostatic challenge paradigm, the increase in plasma NE produced by standing up is consistently greater in depressed unipolar **or** bipolar patients than in age and sex-matched controls (Siever, 1987; Manji et al., 1994). Depressed patients also show significantly higher plasma NE levels following a cold pressor test than age- and sex-matched controls, compatible with a dysregulation of peripheral release of NE in affective illness. The CSF studies of NE and its metabolite

MHPG suggest that NE output is higher in mania than in depression and that there may be relatively higher values in unipolar versus bipolar depression. Relative elevation within patient groups may, in turn, be related to anxiety or the overall severity of the condition. It is possible that the CSF findings reflect events occurring in the sympathetic nervous system as much as events occurring in the brain, and it is therefore not surprising that the pattern of findings is similar in CSF and plasma.

Overall, considerable evidence suggests that depressed patients excrete disproportionately greater amounts of NE and its major extraneuronal metabolite, normetanephrine (NMN) relative to total catecholamine synthesis compared to controls. Findings of increased fractional urinary output of urinary NE and NMN and of an exaggerated raise in plasma NE upon orthostatic challenge in depressed patients are compatible with dysregulation of NE release from presynaptic terminals (Potter and Manji, 1994; Manji et al., 1995). The mechanism(s) for this remains unclear but may involve both central and peripheral sites. One possible mechanism is altered sensitivity of nerve terminal α_2 autoreceptors. As discussed in detail elsewhere, they do not directly influence the firing rate of the neuron but attenuate the release of neurotransmitters when an action potential depolarizes the varicosity or terminal or may even prevent the action potential from "invading" the nerve terminal (Starke et al., 1989). Altered sensitivity of peripheral α_2 autoreceptors might explain the greater fractional excretion of NE and an exaggerated NE release upon any activation of the sympathetic nervous system (e.g., orthostatic stress, cold stress, "early hospitalization stress", etc.). Given this substantial body of indirect evidence for alterations in α_2AR sensitivity in depression, it is not surprising that several studies have attempted to study α_2ARs more directly.

PLATELET α_2 RECEPTORS IN DEPRESSION

Due to the accessibility of platelets, α_2ARs have been most extensively studied. Any assumption, however, that changes in adrenergic receptors on peripheral cells reflect similar alterations in the CNS is probably not valid, since blood cells are, by definition, non-innervated, exist in a markedly different environment, and may therefore poorly reflect central, innervated, adrenergic receptors. Nonetheless, any identified differences between patients and controls may provide findings that could be ultimately shown to reflect important state or trait alterations in the noradrenergic system, considered as a whole (that is neurotransmitter release, metabolism, receptor density and regulation). Data generated from radioligand binding studies in human platelets most commonly involve ^{3}H-yohimbine as a selective α_2AR antagonist. The overall data using yohimbine generally does not yield significant differences between depressed patients and control groups for B_{max} or K_d (reviewed in Piletz et al., 1986; Kafka and Paul, 1986; Horton et al., 1986; Grossman et al., 1992). The range of mean values for B_{max} and K_d in depressed patients and control groups varies 3–4 fold and 4–5 fold respectively. Some groups attributed the absence of positive findings, with yohimbine as an α_2AR antagonist, to its binding characteristics (reviewed in Piletz et al., 1986; Kafka and Paul, 1986; Horton et al., 1986; Grossman et al., 1992), emphasizing that an antagonist cannot reflect the properties of receptors in the high affinity state which may be more relevant to alterations in norepinephrine concentrations. Radioligand binding studies across groups are confounded

TABLE 1
Adrenergic receptors and responsivity in depression.

Experimental Parameter	Depressed vs Control
Platelet α_2 Receptors	
$\quad B_{max}$ – Antagonist (Yohimbine) Binding	No Change
$\quad B_{max}$ – Agonist (Clonidine) Binding	Increased
$\quad$ Function	? Subsensitive
Lymphocyte β_2 Receptors	
$\quad B_{max}$ – Antagonist Binding	Inconsistent
$\quad$ Function – cAMP Response to Isoproterenol	Subsensitive
Neuroendocrine Responses	
$\quad \alpha_2$ Agonist (Clonidine): MHPG Response	No Change
$\quad\quad$ ACTH & Cortisol	Inconsistent
$\quad\quad$ Growth Hormone	Decreased
$\quad \alpha_2$ Antagonist (Yohimbine): MHPG Response	No Change

by methodological problems, such as varying patient composition, sex, age, clinical state, drug washout period, assay and platelet harvesting techniques, all of which have been extensively reviewed (Piletz *et al.*, 1986; Kafka and Paul, 1986; Horton *et al.*, 1986; Grossman *et al.*, 1992; Piletz *et al.*, 1994).

By contrast, the overall data using clonidine as a radioligand for α_2ARs on platelet membranes yield substantially different and more consistent results. Of seven studies reporting values for depressed patients and control groups, six report an **increase** in the mean values of B_{max} in depressed patients (reviewed in Piletz *et al.*, 1986; Kafka and Paul, 1986; Horton *et al.*, 1986; Grossman *et al.*, 1992; Piletz *et al.*, 1994). These results have generally been considered to be supportive of the α_2 supersensitivity hypothesis of depression, but there are limitations to interpreting findings with clonidine that go beyond reproducibility of measures across groups. First and foremost is the recognition that clonidine binds to (imidazoline) sites that do not belong to any class of α_2ARs (Michel *et al.*, 1989), and the differences in the results using clonidine or yohimbine as the radioligand may reflect differences in the density of platelet imidazoline receptors rather than α_2ARs (Piletz *et al.*, 1994). Additionally, the functional studies of platelet α_2AR (that is inhibition of adenylyl cyclase) tend to suggest *subsensitive* α_2ARs in depression (see Table 1). Similarly, although the overall evidence from pharmacologic challenge studies in depression is inconsistent (discussed below), most studies find a blunted growth hormone (GH) response to clonidine in depression.

α_2AR "CHALLENGE" STUDIES IN DEPRESSION

Pharmacologic challenge paradigms utilize agents known to directly or indirectly stimulate receptor sites and have been extensively employed for testing pathophysiological hypotheses about noradrenergic dysfunction in affective illness (Siever, 1987; Manji *et al.*, 1995). The α_2AR partial agonist clonidine (which also exerts effects at imidazoline sites, see above) has been administered to depressed patients, and changes in plasma MHPG, blood pressure, heart rate, sedation, and/or cortisol have been measured; overall,

TABLE 2

α_2 adrenergic receptors and mood disorders.

Blunted GH response to clonidine in depressed patients

Exaggerated NE response to orthostatic challenge in unipolar and bipolar patients

Elevated ^{3}H clonidine binding to platelet membranes from depressed patients

Subsensitivity of rat brain α_2 adrenergic receptors after chronic treatment with antidepressants (TCAs, MAOIs, ECS)

Subsensitivity of rat brain and human platelet α_2 adrenergic receptors after chronic treatment with lithium

Antidepressant efficacy of compounds with α_2 antagonistic properties: mianserin, idazoxan and mitrazapine

Reduced α_2 inhibition of adenylyl cyclase in platelets from depressed patients

Elevated ^{3}H clonidine binding in frontal cortex from tissue obtained at post-mortem from suicide victims

---

the results have been inconsistent. In contrast, almost all studies show a significantly reduced GH response to clonidine (presumably mediated by post synaptic hypothalamic α_2 receptors) in depressed patients (Siever, 1987; Potter and Manji, 1994). However, it is clear that this response is not specific for depression, since a blunted GH response to clonidine has been reported in patients with panic disorder, generalized anxiety symptoms, obsessive-compulsive disorder, and even in mania (Manji et al., 1995). Thus, a blunted α_2AR response may be observed in several conditions all of which may involve tonic or episodic elevations of NE in the brain.

RATIONALE FOR THE USE OF α_2AR ANTAGONISTS IN THE TREATMENT OF DEPRESSION

Given the uncertainty about the pathophysiologic basis of any abnormalities of noradrenergic systems in depression and α_2ARs in particular, there is no *direct* argument for a therapeutic effect of α_2AR antagonism in depression. Three indirect pharmacologic rationales, however, have led investigators to seriously consider such a possibility (see Table 2). First, the more traditional formulation of how altered α_2ARs may play a role in depression is linked to findings that the therapeutic effect of chronic antidepressant treatments may result from a decrease in the number of central presynaptic α_2ARs. The proposed result would be less inhibition of norepinephrine release and a corresponding relative increase in intrasynaptic norepinephrine (reviewed by Kafka and Paul, 1986). Secondly, as has been reviewed elsewhere (Pinder and Seizin, 1987; Dickinson, 1991), it has been speculated that the antidepressant effects of mianserin are related to its antagonism of α_2ARs, mainly because no other mechanism was identified. Finally, the combination of α_2AR antagonists and norepinephrine uptake inhibitors leads to a greatly accelerated time course of β_1-adrenoceptor down regulation in rat brain (Crews et al., 1981; Keith et al., 1986). Since, β-adrenoceptor down regulation has been argued to be one of and perhaps the most common chronic biochemical effect of antidepressant treatments (Sulser et al., 1978; Manji et al., 1991), the potentiating effect of α_2 antagonism is highly suggestive. This has led to the speculation that not only might α_2AR antagonists be efficacious antidepressants alone, they might also potentiate the therapeutic actions and/or accelerate clinical effectiveness of "traditional" antidepressants (Dickinson, 1991). Since the effects of NE reuptake inhibitors and MAOIs depends on the existing "firing rate" of NE neurons, it has been postulated that α_2AR antagonists may even be effective in patients who have failed to respond to "traditional" ADs. This is because α_2AR antagonists should not only possess the ability to enhance the release of transmitter at the terminal, but also to increase the rate of cell firing by blocking somatodendritic α_2ARs in the major CNS noradrenergic neurons.

α_2AR ANTAGONISTS AS ANTIDEPRESSANTS

As has already been discussed, the antidepressant effects of mianserin may be related to its antagonism of α_2ARs. The first attempt to directly test the antidepressant efficacy of α_2AR antagonists was to add yohimbine to tricyclic antidepressants, generally in patients who had not responded to the tricyclic alone (Charney et al., 1986). In this group of refractory patients, the addition of yohimbine was without benefit. By contrast, one study has demonstrated a "dramatic reponse" to electroconvulsive therapy in three subjects pretreated with yohimbine (Sacks et al., 1986). Although results using yohimbine are discouraging (Dickinson, 1991), our studies using idazoxan, which is a more specific agent, have been encouraging.

STUDIES OF IDAZOXAN IN HEALTHY VOLUNTEERS

Chronic physiological, biochemical, and behavioral effects from administration of putative α_2AR antagonists are difficult to predict since most preclinical and clinical studies have been limited to measuring effects within the first 2–3 h following administration (Potter et al., 1994). For instance, there are a number of studies in rats which demonstrate 'positive' responses to idazoxan and/or yohimbine in behavioral paradigms that are distinguishable from those of amphetamine but can be broadly classified as stimulatory or enhancing some aspects of retention of learned behaviors. Clinical investigations have also shown that acute yohimbine, idazoxan, and atipamezole produce variable increases from 100–600 per cent of, for instance, plasma norepinephrine that are maximal by 20–30 min following intravenous infusion and return to baseline by 90 min (reviewed in Potter et al., 1994). From the available data it is impossible to assess whether differences in the magnitude of increase of plasma norepinephrine (greatest after atipamezole) have more to do with differences in the specificity, potency and/or pharmacokinetics of the compounds. Acute increases of blood pressure are considered to be dose limiting so that full dose-response curves to establish the maximal norepinephrine reponse to each dose are not feasible in humans. Although of uncertain relevance to effects on peripheral catecholamines, idazoxan is reported to produce changes in the baseline EEG of awake rats that are not seen after the more selective α_2AR antagonists, efaroxan and ethoxyidazoxan (Dickinson and Gadie, 1991). Such data may, however, be relevant to reports of differential effects of bedtime administration of 150 mg maprotiline, 0.10 mg clonidine, and 40 mg idazoxan per os (Nicholson and Pascoe, 1991).

To date, published clinical studies on biochemical, physiologic, and behavioral effects following chronic administration of α_2AR antagonists are only available for idazoxan in healthy volunteers. In an unusually comprehensive study, idazoxan was administered at a dose of 120 mg/day for three weeks to 12 normal male volunteers to assess effects on cortisol and ACTH, plasma 3-methoxy-4-hydroxy-phenylglycol (MHPG), melatonin, platelet and lymphocyte adrenoceptor binding (Glue et al., 1991, 1992). Chronic idazoxan reduced plasma MHPG while having no effect on nocturnal plasma melatonin, diastolic blood pressure, or heart rate (Glue et al., 1991). Interestingly, there was a tendency for chronic idazoxan, opposite to its acute effects, to reduce baseline (pre-AM dose) resting values systolic blood pressure, suggesting 'an upregulation of central inhibitory α_2ARs and subsequent reduction in noradrenergic activity' (Glue et al., 1991). In these same subjects, chronic idazoxan tended to reduce total sleep and produced a significant decrease in REM sleep time accompanied by a marked increase in REM latency, a pattern of effects noted to be most similar to those observed after the MAOI class of antidepressants (Potter et al., 1994). Second, chronic but not acute idazoxan increased the density of platelet α_2AR number (Glue et al., 1991). Taken together, the results from these and other studies are consistent with acute central and peripheral noradrenergic effects of idazoxan followed by compensatory chronic changes as reflected in reduced total norepinephrine turnover (decreased MHPG), α_2AR changes and a return of both cortisol and cardiovascular regulation to normal (or below).

We have recently completed a protocol involving administration of 120 mg/day of idazoxan to healthy volunteers for 2-week period, looking at a somewhat different group of measures. We did not observe any gross behavioral or cardiovascular changes above pretreatment levels, findings consistent with the observations of Glue et al. (1991) described above. Taken together, studies of idazoxan in healthy volunteers show that, at doses producing subtle chronic alterations of noradrenergic systems, remarkably few physiologic or behavioral effects are detectable. In fact, our subjects who received idazoxan under blind conditions could not detect above chance whether they were on drugs, a finding consistent with the absence of behavioral effects on self-rated visual analog scales. Indeed, the only parameter tending to change was baseline anxiety which was in the direction of a decrease (Potter et al., 1994). This would be the expected direction of change over time in anxiety for volunteers participating in a research protocol involving repetition of procedures to which the subjects became more accustomed.

α_2AR ANTAGONISTS IN THE TREATMENT OF DEPRESSION

We previously reported our initial positive experience with idazoxan in a double-blind, placebo-controlled trial in treatment-resistant, severely depressed patients. Interestingly, two of these patients had previously been unresponsive to several conventional treatments, including tricyclics, MAOIs, lithium carbonate and ECT (Osman et al., 1989). Over the last 3 years we have carried out a 6–8 week trial of 80–120 mg/day of idazoxan alone or added to a stable dose of lithium in a mixed population of 13 bipolar I, II and unipolar patients, as well as one not otherwise classifiable depressed patient (Table 3). All patients received idazoxan under double-blind conditions during extended (minimum of four months) hospitalization. A positive clinical response was judged to be present if there was

TABLE 3
Characteristics of 4–6 week inpatient idazoxan trial in treatment-resistant depression.

Pt.	Sex	Dx	Response
1	F	BPI	No
2	M	UP	No
3	F	UP	Yes (with panic attack)
4	M	BPI	Yes
5	F	BPII	Yes
6	M	BPI	Prohylactic
7	F	UP	No
8	F	BPII	Yes
9	F	BPII	Yes
10	F	UP	No
11	F	UP	No
12	F	UP	No
13	M	NOS	No

BPI, Bipolar I (depression plus mania); BPII, Bipolar II (depression plus hypomania); UP, Unipolar (depression only).

a three-point or greater reduction in the daily Bunney-Hamburg (B-H; Bunney and Hamburg, 1963) rating comparing the average of daily ratings for the week prior to beginning treatment to the average values following 4–6 weeks of treatment. For reference purposes, a total depression score of 6 on the B-H corresponds to a HDRS score of 25. A value of 2 or less on the B-H corresponds to remission (HDRS < 8). To date, as shown in Table 3, six patients (if one includes an apparent prophylactic response in subject No. 6) have responded. What was striking to us is that only 1/7 unipolar or nonbipolar patients responded versus 5/6 bipolars. We have therefore just initiated a prospective trial of idazoxan in bipolar patients among whom we will maintain those with a history of recurrent mania on lithium.

Our initial analysis of biochemical effects of idazoxan in the first six to ten of the patients listed in Table 4 provides evidence for compensatory changes in with our initial observation in three patients, we observe increases of lying (mean ± SD pre-idazoxan 0.79 ± 0.48 pmol/ml; post-idazoxan 1.67 ± 0.71 pmol/ml) and standing plasma norepinephrine

TABLE 4
Effects of chronic idazoxan (4–6 weeks) on urinary catecholamines in depressed patients.

	Pre vs. Post $\mu M/24\ h \pm SD$		p value
Norepinephrine (NE)	0.67 ± 0.22	0.68 ± 0.18	> 0.10
Metanephrine	0.45 ± 0.18	0.51 ± 0.18	0.08
Normetanephrine (NM)	1.42 ± 0.60	1.61 ± 0.62	> 0.10
MHPG	9.78 ± 4.77	7.90 ± 3.10	0.07
VMA	14.8 ± 5.5	12.4 ± 3.6	0.06
HVA	21.0 ± 4.2	23.0 ± 6.3	> 0.10
NE + NM/MHPG + VMA	0.086 ± 0.014	0.115 ± 0.018	< 0.001
HVA/MHPG + VMA	0.95 ± 0.36	1.14 ± 0.28	< 0.015

MHPG, 3-methoxy-4-hydroxyphenylglycol; VMA, vanillylmandelic acid; HVA, homovanillic acid.

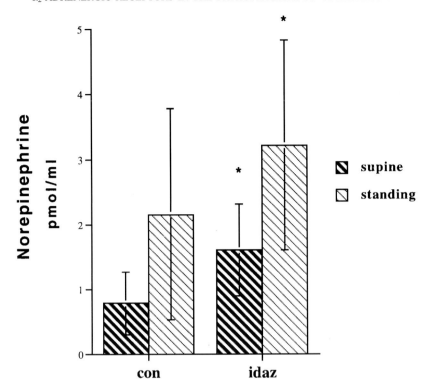

FIGURE 1 Effects of chronic idazoxan on plasma norepinephrine in depressed subjects. 10 depressed patients were administered idazoxan chronically (3–6 weeks), with or without concomitant lithium carbonate administration. Subjects were free of all medications (with the possible exception of lithium carbonate) for at least 3 weeks before beginning the study. Supine samples were obtained following an overnight fast, between 8:00 and 9:00 AM, from an antecubital indwelling catheter, at least 45 minutes after needle insertion. Standing samples were collected after the subject had been standing for 5 minutes. Plasma was assayed for norepinephrine using high-performance liquid chromatography with electrochemical detection. *$p < 0.05$, paired t-test.

(pre 2.16 ± 1.6 pmol/ml; post 3.22 ± 1.61 pmol/ml) (Figure 1). And, consistent with Glue et al.'s (1991) report of decreased free plasma MHPG in volunteers after idazoxan, total (free plus conjugated) MHPG tended ($p < 0.07$) to fall in urine (Table 4). What was most striking in the urinary measurements was that the ratio of norepinephrine + normetanephrine to MHPG + VMA increased in all but one subject investigated (Table 4 and Figure 2). This ratio can be considered as a rough index of "extraneuronal" norepinephrine since essential steps in the formation of MHPG and vanillylmandelic acid, which depend on deamination, occur primarily intraneuronally (Kopin, 1985). The most obvious interpretation of the biochemical data in patients is that chronic blockade of α_2ARs produced a modest decrease in centrally mediated sympathetic nervous system while producing a relative enhancement of release of sympathetic nerve terminal (Manji et al., 1995). The point should be emphasized that we have included patients in the analysis who were on lithium; this may be particularly relevant because lithium is known to exert effects on the inhibitory G protein G_i (Risby et al., 1991), effects which may result in an uncoupling of α_2ARs from signal transduction pathways.

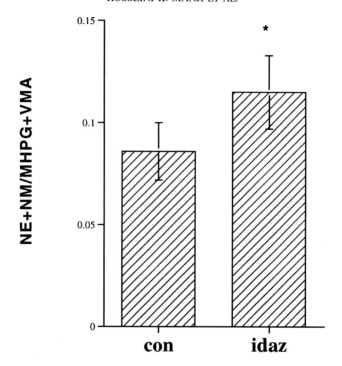

FIGURE 2 Effects of chronic idazoxan on the ratio of extraneuronal norepinephrine urinary metabolites in depressed subjects. 10 depressed patients were administered idazoxan chronically (3–6 weeks), with or without concomitant lithium carbonate administration. Subjects were free of all medications (with the possible exception of lithium carbonate) for at least 3 weeks before beginning the study. All subjects were maintained on a low monoamine, low caffeine diet, and two consecutive 24 hr urine samples were collected. The 24 hr urine was collected into refrigerated plastic containers containing 3% sodium metabisulfite as an antioxidant. Urine was assayed for total (free plus conjugated) NE, MHPG, VMA, and NMN using gas chromatography with mass fragmentographic detection methods. *$p < 0.05$, paired t-test.

CONCLUDING REMARKS

To date, clinical studies of overall noradrenergic function, attempts to measure α_2AR function and initial therapeutic trials in depressed are strongly suggestive of some abnormality which can be beneficially modified by administration of α_2AR antagonists. None of the measures obtained can, *with certainty*, be attributed to alteration or manipulation of some specific class of α_2AR. Much of the data, however, is consistent with such a possibility. Moreover, it is not clear that, with doses and schedules of administration so far tested, maximal tolerated α_2AR blockade in the central nervous system has been produced. Nevertheless, idazoxan has been demonstrated as an effective antidepressant with remarkably few side effects, even in patients who are resistant to established therapies. Most recently, there have been studies with the 6-azaanalogue of mianserin, mirtazepine (Org3770), which also possesses α_2AR antagonist properties (deBoer and Ruigt, 1995). To date, this agent appears to be a very effective agent, suggesting that α_2AR antagonists may represent a much needed addition to the therapeutic armamentarium in the treatment of depression.

References

Bunney, W.E. Jr and Hamburg, D.A. (1963). Methods for reliable longitudinal observation of behavior. *Archives of General Psychiatry*, **9**, 114–28.

Charney, D.S., Price, L.H. and Heninger, G.R. (1986). Desipramine-yohimbine combination treatment of refractory depression. Implications for the β-adrenergic receptor hypothesis of antidepressant action. *Arch. Gen. Psychiatr.*, **43**, 1155–61.

DeBoer, T.H. and Ruigt, G.S.F. (1995). The selective α_2-adrenoceptor antagonist mitrazapine (Org3770) enhances noradrenergic and 5-HT-1A-mediated serotonergic neurotransmission. *CNS Drugs*, **4** suppl, 29–38.

Doxey, J.C., Roach, A.G. and Smith, C.F.C. (1983). Studies on RX781094: A selective, potent and specific antagonist of α_2-adrenoceptors. *Brit. J. Pharmacol.*, **78**, 489–505.

Elliott, H.L., Jones, C.R., Vincent, J., Lawrie, C.B. and Reid, J.L. (1984). The α adrenoceptor antagonist properties of idazoxan in normal subjects. *Clin. Pharmacol. Ther.*, **36**, 190–6.

Glue, P., Payvandi, N., Kay, G., Elliott, J.M. and Nutt, D.J. (1991). Effects of chronic α_2-adrenoceptor blockade of platelet and lymphocyte adrenoceptor binding in normal volunteers. *Life Sciences*, **49**, 21–5.

Glue, P., Wilson, S., Lawson, C., Campling, G.M., Franklin. M., Cowen, P.I.J., *et al.* (1992). Acute and chronic idazoxan in normal volunteers: biochemical, physiological and psychological effects. *Journal of Psychopharmacology*, **5**, 394–401.

Grossman, F., Manji, H.K. and Potter, W.Z. (1993). Platelet α_2 adrenoreceptors in depression: a critical examination. *Journal of Psychopharmacology*, **7**, 4–18.

Horton, R.W., Katona, C.L.E., Theodorou, A.E., Hale, A.S., Davies, S.L., Tunnicliffe, C., *et al.* (1986). Platelet radioligand binding and neuroendocrine challenge tests in depression, antidepressant and receptor function. *Ciba Foundation Symposium*, **123**, 84–105.

Kafka, M.S. and Paul, S.M. (1986). Platelet α_2-adrenergic receptors in depression. *Arch. Gen. Psychiat.*, **43**, 91–5.

Keith, R.A., Howe, B.B. and Salama, A.I. (1986). Modulation of peripheral β_1 and α_2 sensitivities by the administration of the tricyclic antidepressant, imipramine, alone and in combination with α_2 antagonist to rats. *Journal of Pharmacology and Experimental Therapeutics*, **236**, 356–63.

Kopin, I.J. (1985). Catecholamine metabolism: basic aspects and clinical significance. *Pharmacological Reviews*, **37**, 333–64.

Manji, H.K., Bitran, J.A., Chen, G., Gusovsky, F. and Potter, W.Z. (1992). Idazoxan downregulates β adrenergic receptors on C6 glioma cells *in vitro*. *European Journal Pharmacology: Molecular Pharmacology Section*, **227**, 275–82.

Manji, H.K., Rudorfer, M.V. and Potter, W.Z. (1994). Affective disorders and adrenergic function. In O.G. Cameron (ed.) *Adrenergic Dysfunction and Psychobiology*, pp. 365–401. American Psychiatric Press. Washington.

Manji, H.K. and Potter, W.Z. (1995). Affective Disorders. In Pullan, L. and Patel, J. (eds.) *Emerging Strategies in Neurotherapeutics*, Humana Press, in press.

Michel, M.D., Regan, J.W., Gerhardt, M.A., Neubig, R.R., Insel, P.A. and Motulsky, H.J. (1990). Nonadrenergic [^{3}H]idazoxan binding sites are physically distinct from α_2-adrenergic receptors. *Molecular Pharmacology*, **37**, 65–8.

Osman, O.T., Rudorfer, M.V., Manji, H. and Potter, W.Z. (1989). Idazoxan: A novel α_2-antagonist antidepressant. *142nd APA Meeting* (May, San Francisco), Abst.

Piletz, J.E., Halaris, A. and Ernsberger P.R. (1994). Psychopharmacology of imidazoline and α_2-adrenergic receptors: Implications for depression. *Critical Reviews in Neurobiology*, **9**, 29–66.

Piletz, J.E., Schubert, D.S. and Halaris, A. (1986). Evaluation of studies on platelet α_2 adrenoceptors in depressive illness. *Life Sci.*, **39**, 1589–616.

Pinder, R.M. and Sitsen, J.M.A. (1987). α_2-adrenoceptor antagonists as antidepressants: The search for selectivity. In Dahl, Gram, Paul and Potter (Eds.), *Clinical Pharmacology in Psychiatry*, pp. 107–12. Springer-Verlag, Berlin.

Potter, W.Z., Manji, H.K., Osman, O.T. and Rudorfer, M.V. (1992). New antidepressants and their possible modes of action. In I.J.P. Macher & M.A. Crocq (Eds.), *New Prospects in Psychiatry. The Bio-Clinical Interface*, pp. 113–129. Elsevier Science Publishers. New York.

Potter, W.Z., Grossman, F., Dawkins, K. and Manji, H.K. (1994). Initial clinical psychopharmacological studies of α_2-adrenoceptor antagonists in volunteers and depressed patients. In S.A. Montgomery and T.H. Corn (Eds.) *Psychopharmacology of Depression*. Oxford University Press.

Potter, W.Z. and Manji, H.K. (1994). Affective disorders and adrenergic function: an update. *Clinical Biochemistry*, **40**, 279–287.

Rudorfer, M.V., Manji, H.K. and Potter, W.Z. (1994). Comparative tolerability profiles of the newer versus older antidepressants. *Drug Safety*, **10**, 18–46.

Risby, E.D., Hsiao, J.K., Manji, H.K., Bitran, J., Moses, F., Zhou, D.F., *et al.* (1991). The mechanisms of action of lithium. II. Effects on adenylate cyclase and β receptor bindings in normals. *Archives of General Psychiatry*, **48**, 513–24.

Sachs, G.S., Pollack, M.H., Brotman, A.W., Farhadi, A.M. and Gelenberg, A.J. (1986). Enhancement of ECT benefit by yohimbine. *J. Clin. Psychiatr.*, **47**, 508–10.

Scott, J.A. and Crews, F.T. (1983). Rapid decrease in rat brain β adrenergic recepor binding during combined antidepressant α_2 antagonist treatment. *J. Pharmacol. Exp. Ther.*, **224**, 640–6.

Siever, L.J. (1987). Role of noradrenergic mechanisms in the etiology of the affective disorders. In H.Y. Meltzer (Ed.), *Psychopharmacology, the Third Generation of Progress*, pp. 493–504. Raven Press, New York.

Starke, K., Gothert, M. and Kilbinger, H. (1989). Modulation of neurotransmitter release by presynaptic autoreceptors. *Physiological Reviews*, **69**, 864–989.

Sulser, F., Vetulani, J. and Mobley, P.L. (1978). Mode of action of antidepressant drugs. *Biochemical Pharmacology*, **27**, 257–71.

CLINICAL APPLICATIONS OF α_2 ADRENERGIC AGONISTS IN THE PERIOPERATIVE PERIOD — NEUROBIOLOGIC CONSIDERATIONS

MERVYN MAZE and TOSHIKI MIZOBE

Department of Anesthesia, Stanford University School of Medicine, Palo Alto CA 94305, USA

α_2 adrenergic agonists are efficacious for anxiolysis, preoperative sedation, and for decreasing anesthetic requirements for volatile, narcotic and anesthetic induction agents. Additionally, α_2 adrenergic agonists minimize hemodynamic instability, facilitate hypotensive techniques, decrease postoperative shivering and oxygen utilization and are effective at protecting against ischemia of the heart and brain. Because of the perceived benefits of these compounds in the surgical patient's care, these have now been advocated for use over the entire perioperative period.

However, following chronic α_2 adrenergic agonist administration in rats, tolerance rapidly develops to the sedative effects through an uncoupling of the receptor to its effector in the locus coeruleus. Tolerance develops more slowly to the analgesic and anesthetic sparing effects of α_2 agonists. The sympatholytic action of α_2 agonists do not desensitize following chronic administration, even at very high doses. The differential development of tolerance for the different responses relates to the efficiency of transduction. More than 80% of the receptors have to be available for sedation, 40% for analgesia, 20% for anesthetic-sparing, and 5% for sympatholysis.

The lack of suitable pharmacologic probes precludes identification of the receptor subtypes by classical pharmacologic approaches. Instead, investigators have resorted to molecular genetic reagents including "knockouts", "knockdowns", and transgenics. Preliminary information with these reagents suggests that most of the responses that occur perioperatively are mediated through $\alpha_{2A/D}$ adrenoceptors. This should be considered as the prime target for drug discovery programs. The serendipitous use of α_2 agonists in the perioperative period has spurred molecular biologic studies to determine (i) which components of the transduction pathways modulate responsiveness, (ii) whether tolerance can be pre-empted, and (iii) the synthesis of novel chemical entities with exquisite sensitivity for the therapeutic target. This area of biomedical investigation is illustrative of the rapid transfer of cutting edge molecular technology to clinical care.

KEY WORDS: adrenergic receptors, α_2, agonists, dexmedetomidine, anesthesia, analgesia, desensitization, molecular modeling, gene-targeting, antisense

INTRODUCTION

Clonidine, an α_2 adrenoceptor agonist (α_2 agonist), was introduced into clinical practice as an antihypertensive medication more than 25 years ago. Apart from its use as an antihypertensive medication, clonidine has been used for many conditions ranging from a variety of psychiatric disorders to the management of children with delayed growth. In veterinary practice, α_2 agonists have been used for several years in the setting of anesthesia. Experimental and clinical studies have now progressed to the stage that human

---

Correspondence to: Dr Mervyn Maze, Dept. of Anesthesia, Stanford University School of Medicine, VAPAHCS Palo Alto CA 94305, USA. Tel: 415-493-5000 ext 64224 Fax: 415-852-3417

anesthesiologists are focusing on the use of this class of agent for its analgesic and anesthetic effects. This renewed interest has coincided with the development and clinical introduction of super-selective agonists such as dexmedetomidine.

PHARMACOLOGICAL RESPONSES OF α_2 AGONISTS ON DIFFERENT SYSTEMS

Central nervous system

Sedation is one of the most consistent effects mediated by central α_2 receptors. While this property is an undesirable side-effect when clonidine is administered for patients with hypertension, it has been used to great advantage for a premedicant in anesthesia. This effect of α_2 agonists is significantly potentiated when administered together with a benzodiazepine. Recently (Correa-Sales et al., 1992a), the locus coeruleus was shown to be a principal region responsible for the sedative effect. The molecular components which participate in the signal transduction of the hypnotic response to α_2 agonists include a postsynaptic α_2 adrenoceptor (Segal et al., 1988) and a pertussis toxin-sensitive G protein (Correa-Sales et al., 1992b) coupled to specific ion channels (Nacif-Coelho et al., 1994). Another characteristic effect of α_2 agonists in clinical situations is anxiolysis, which is comparable to that produced by benzodiazepines compounds. Clonidine can also depress panic disorder in humans. However, higher doses of α_2 agonists may produce an anxiogenic response through non-selective activation of α_1 receptors.

α_2 adrenergic receptor activation produces a potent analgesic response, involving both supraspinal and spinal sites. In animal experiments clonidine exerts a more potent analgesic effect than that exerted by morphine. Furthermore, the analgesic potency of α_2 agonists is synergistically enhanced with concomitant treatment with opioids. Combining clonidine with opiate narcotics will lead to a lower dose requirement for each drug while reducing the incidence and severity of side effects. α_2 agonists suppress the undesirable physiological and psychological symptoms following withdrawal from opiates. Recently, the usefulness of α_2 agonists has been extended to other withdrawal states, such as alcohol, benzodiazepine, and even nicotine "craving". In humans, dexmedetomidine was reported to suppress ischemic pain and to attenuate the affective component of ischemic pain; yet, intravenous dexmedetomidine, in the dose range of 0.25 and 0.50 μg/kg, did not affect the experimental pain threshold.

The most impressive action of α_2 agonists in the central nervous system is its ability to reduce anesthetic requirements. A modest reduction (15%) of halothane MAC follows subacute administration of clonidine in rabbits. Acutely-administered clonidine reduced halothane MAC by 50% in a dose-dependent fashion; this MAC-reducing effect was antagonized by an α_2 antagonists. There is a ceiling to the MAC-reducing properties of clonidine due to the drug's affinity for, and activation of, α_1 adrenoceptors at higher concentrations. Activaton of α_1 adrenoceptors will functionally antagonize α_2 agonist action in the central nervous system (Guo et al., 1991). More selective α_2 agonists are able to reduce the MAC of volatile anesthetics to a much greater extent. Azepexole was shown to reduce isoflurane MAC by 85% in dogs, while dexmedetomidine, the most selective α_2 agonist, decreased halothane MAC by more than 95% in animals, indicating that it alone may produce the anesthetic state (Vickery et al., 1988). The opiate receptor is not involved in the anesthetic-sparing action of α_2 agonists. This reduction in anesthetic requirement can also be demonstrated in humans and is not limited to volatile anesthetics (see below).

Intraocular pressure can be reduced by α_2 agonists and these agents can also attenuate the rise in intraocular pressure elevation associated with laryngoscopy and endotracheal intubation. Whether through the same, or different mechanisms, raised intracranial pressure is decreased by α_2 agonists in an animal model of subarachnoid hemorrhage. Experimental application of α_2 agonists and antagonists in the study of neuroprotection from cerebral ischemia has resulted in conflicting data. Hoffman *et al.* reported that the α_2 agonists, clonidine and dexmedetomidine, could improve outcome from incomplete global ischemia. Recently, the neuroprotective effect of dexmedetomidine has been reported in a rabbit model of focal ischemia in which a beneficial effect was demonstrated even when the α_2 agonist was administered *after* the onset of ischemia (Maier *et al.*, 1993). On the other hand, idazoxan, an α_2 antagonist, also protects against global ischemia. This apparent paradox may be reconciled by the finding that both idazoxan and rilmenidine, α_2 antagonist and agonists with affinity for the imidazole-preferring receptor, are able to exert a protective against cerebral ischemia. It is hypothesized that the imidazoline-preferring receptor, and not α_2 adrenoceptors, are involved in the neuroprotective mechanism. Whichever receptor mediates the neuroprotective effect it is unlikely to be due to a vascular action since dexmedetomidine has been shown to decrease cerebral blood velocity in humans in a dose-dependent manner (Zornow *et al.*, 1993).

Cardiovascular system

The action of α_2 agonists on the cardiovascular system may be classified as peripheral or central. α_2 agonists inhibit norepinephrine release from peripheral prejunctional nerve endings and this property, in part, contributes to the bradycardic effect of α_2 agonists. Until now there is no firm evidence to support the existence of postsynaptic α_2 receptors in the myocardium; therefore, it is unlike that the α_2 agonists exert a direct effect on the heart. Postjunctional α_2 receptors are present in both arterial and venous vasculature where they produce vasoconstriction. Among the different vascular beds the effect of α_2 agonists in the coronary circulation is important from the clinical standpoint. A putative vasoconstrictive action of α_2 agonists on the coronary vasculature could promote ischemia; however, these agonists can ameliorate any direct constriction by reducing sympathetic outflow. Furthermore, α_2 agonists also have been documented to release endothelial-derived relaxant factor (nitric oxide) in coronary arteries and to enhance coronary blood flow induced by endogenous and exogenous adenosine in an *in vivo* model. Thus, the effect on coronary arteries may be too complex to distinguish clear changes in coronary blood flow in an *in vivo* model. Intrathecal administration of clonidine shows a biphasic effect on blood pressure with a small dose (150 µg) inducing hypotension while a larger dose (450 µg) causes hypertension presumably due to peripheral vasoconstriction. An intermediate dose (300 µg) has little effect on blood pressure presumably due to offsetting peripheral and central effects.

Clonidine can produce hypotensive and bradycardic effects *via* a site in the central nervous system. The mechanism for these actions may involve inhibition of sympathetic outflow and potentiation of parasympathetic nervous activity. However, the precise mechanism involved in these actions are not well understood. While the nucleus tractus solitarius (a site known to modulate autonomic control including vagal activity) is an important central site for the action of α_2 agonists, other nuclei including the locus coeruleus, the dorsal motor nucleus of vagus and the nucleus reticulalis lateralis may also mediate hypotension and/or bradycardia. Bradycardia is especially likely to occur in patients with little counteracting sympathetic stimulation. Recently, the imidazoline-preferring receptors were shown to play an important role in the hypotensive effect of α_2 agonist. They also suggested that the α_2 agonists exert their hypotensive and sedative effects in different receptor-effector mechanisms, respectively.

Another cardiovascular property of α_2 agonists is its anti-arrhythmic effect. Dexmedetomidine attenuates epinephrine-induced arrhythmias during halothane anesthesia and also the arrhythmias that occur from bupivacaine toxicity. The anti-arrhythmic action in halothane-epinephrine arrhythmias is abolished in vagotomized animals suggesting that enhanced vagal activity mediates this action.

The effect of α_2 agonist on the cerebral circulation during anesthesia has been studied. Dexmedetomidine decreases cerebral blood flow in awake human volunteers (Zornow *et al.*, 1993). This characteristic may be favorable in protecting the brain from an abrupt increase of blood flow. This idea has been supported by a recent finding that dexmedetomidine blunts the cerebrovascular response to hypoxic hypoxia during isoflurane anesthesia.

Respiratory system

The respiratory depressant effects of clonidine are not remarkable unless massive doses are given. Intravenous clonidine induces hypoxia in ungulates due to platelet aggregation; this effect is not found in humans. Although α_2 agonist may cause mild respiratory depression, the effect of clonidine is less than that of opiate narcotics. In clinically appropriate doses, mild respiratory depression can be detected with hypercarbic ventilatory response studies. This appears to be equivalent to the respiratory depression that can be detected with physiologic sleep. There is no potentiation by clonidine of opioid-induced respiratory depression. Nebulized clonidine was demonstrated to attenuate bronchoconstriction in asthmatic patients and has also been prescribed to patients with obstructive sleep-apnea syndrome.

Endocrine system

The α_2 agonists potentiate the secretion of growth hormone. Although a precise mechanism for this action has not been elucidated, α_2 receptor activation is coupled to growth hormone releasing factor. α_2 agonists which possess the imidazole ring inhibit steroidogenesis. However, at clinical doses, this mild effect is not likely to have serious consequences. The α_2 agonists decrease sympathoadrenal outflow and these agents can suppress the stress-response following surgical stimulation. While *in vitro* studies indicate that α_2 agonists regulate catecholamine secretion in the adrenal medulla, this effect has been questioned by others. The α_2 agonists also inhibit the release of insulin from the pancreatic β cells directly; again, this action does not result in severe hyperglycemia in clinical settings.

Gastrointestinal system

The α_2 agonists exert a prominent anti-sialogogue effect which is a useful feature in its use as a premedication agent. The α_2 agonists can modulate release of gastric acid via a presynaptic mechanism; yet no significant change in gastric pH is observed in humans. The α_2 agonists also may prevent intestinal ion and water secretion in the large bowel, indicating an effective treatment for watery diarrhoea.

Renal system

The α_2 agonists induce diuresis both in animals and was recently shown to occur in humans. Inhibition of release of antidiuretic hormone (ADH), antagonism of the renal tubular action of ADH, and increase in the glomerular filtration rate have each been implicated in the mechanism. Recently, α_2 agonist-induced release of atrial natriuretic factor has also been suggested to contribute to the diuretic mechanism of α_2 agonists.

Hematologic system

Aggregation of platelets is induced by α_2 agonists. In the clinical setting this is probably offset by the decrease in circulating catecholamines.

USE OF α_2 AGONISTS IN ANESTHETIC PRACTICE

Perioperative administration

Since sedation, anxiolysis and its antisialogogue action are attractive attributes in a premedication agent, administration of α_2 agonists suits this purpose well. The decreased requirement for induction and maintenance anesthetic agents could result in a quicker emergence from anesthesia, although this has yet to be rigorously studied. There are also many clinical reports establishing the efficacy of α_2 agonists both to supplement regional anesthetic techniques and to augment the duration and the quality of conduction anesthesia.

Another route of administration of the α_2 agonists is into intrathecal or epidural space to potentiate local anesthetic agents. Concerning epidural anesthesia, addition of clonidine to epidural lidocaine was reported to augment its anesthetic potency. Epidural clonidine reduces the intra- and postoperative analgesic requirements when compared with the same dose given by the intravenous route, although the plasma concentrations were less in the epidural group.

A potent analgesic property of α_2 agonists provides improved pain control in postoperative patients. Epidural administration is the most common route and has been well investigated. Systemic administration of clonidine for postoperative analgesia has also been reported to be efficacious.

Perioperative as well as postoperative administration of clonidine can decrease oxygen consumption and episodes of shivering during recovery from anesthesia. The anti-shivering effect can be demonstrated with a relaively small dose. This feature provides further justification for using this agent in patients with coronary artery disease.

An extension of the pharmacologic properties of α_2 agonists is the development of severe bradycardia and hypotension. Bradycardia can be pre-empted by the prophylactic use of an anticholinergic agent (atropine or glycopyrrolate). On the other hand hypotension will respond in an exaggerated fashion to treatment with ephedrine.

Use of α_2 agonists in disease states

Two groups of surgical patients in whom the beneficial effects of α_2 agonists have already been demonstrated are those suffering from hypertension and glaucoma. In both groups, the beneficial effects relate to the pharmacologic action of this class of drug to interrupt the pathogenic process. Currently, a number of studies are investigating the potential myocardial ischemic/infarction reducing effect of α_2 agonists including dexmedetomidine and mivazerol. The data with clonidine, while encouraging, are inconclusive. Another group of patients likely to benefit from α_2 agonists are those undergoing neurosurgical procedures because of the putative neuroprotective, ICP-reducing, induced hypotensive, and sympatholytic actions of α_2 agonists.

Miscellaneous usage

Because of a potent analgesic property, α_2 agonists may be useful in the relief of pain other than in the postoperative period. Anecdotal reports demonstrate that intrathecal clonidine in combination with morphine or hydromorphone can attenuate the cancer pain as well as with opioids and the combination was an excellent alternative approach to control of terminal pain. Recently, α_2 agonists have shown promise for intractable pain in patients with reflex sympathetic dystrophy.

TOLERANCE

Chronic administration of dexmedetomidine, a potent and highly-selective α_2 adrenergic agonist, results in the development of tolerance to its hypnotic (Reid *et al.*, 1994) and analgesic responses in rats (Hayashi *et al.*, 1995). In preliminary studies we have investigated the biochemical mechanism for the development of tolerance to the sedative/hypnotic properties (Reid *et al.*, 1996).

The α_2 adrenoceptor affinity (K_i) for displacing ligands (dexmedetomidine, 0.48 ± 0.07 *vs* 0.11 ± 0.03 nM; para-iodoclonidine 2.76 ± 0.48 *vs* 0.63 ± 0.21) was significantly reduced in tolerant rats; B_{max} was unchanged. The ability of PTX to ribosylate G proteins *ex vivo* in tolerant rats was reduced although the quantity of PTX-sensitive G proteins was unchanged. Forskolin-stimulated adenylyl cyclase was less sensitive to inhibition by dexmedetomidine in the tolerant rats. However, acute i.p.-injection of dexmedetomidine still reduced cAMP levels in tolerant rats. The decrease in ribosylation reflects a decrease in the ability of the G protein to couple to the α_2 adrenoceptors in the LC's of tolerant rats. Adrenoceptors in the tolerant rats have a lower binding affinity state probably because of diminished coupling to the G-protein. In this state, the α_2 adrenoceptors are less capable of transducing the effector response (inhibition of adenylyl cyclase).

In preliminary studies we have investigated whether tolerance develops to the sympatholytic and anesthetic-reducing effects of α_2 agonists following prolonged administration of dexmedetomidine and how the number of available α_2 adrenoceptors affects these dexmedetomidine-induced responses (Rabin *et al.*, 1996). The sympatholytic action of dexmedetomidine, assessed neurochemically by the decrease in norepinephrine (NE) turnover, was investigated in the locus coeruleus (LC) and hippocampus. The effect of chronic (7 days) dexmedetomidine (3 and 10 mg·kg^{-1}·h^{-1}) on the drug's sympatholytic action was studied in the same brain regions. The anesthetic-reducing effect of chronic (7 days) dexmedetomidine (5 and 10 mg·kg^{-1}·h^{-1}) was studied by determining the minimum anesthetic concentration (MAC) for halothane which prevented rats from responding to a supramaximal noxious stimulus before and after acute intraperitoneal administration of dexmedetomidine (30 mg·kg^{-1}), a dose which causes a 50% reduction in the MAC for halothane.

The number of α_2 adrenoceptors in the LC, measured by radiolabeled-ligand binding studies, was diminished in a dose-dependent manner by covalent modification with a noncompetitive receptor blocker, N-ethoxycarbonyl-2-ethoxy-1,2-dihydroquinoline (EEDQ). Rats were pretreated with EEDQ (0.3–1.0 mg·kg^{-1}) to delineate the receptor reserve for the NE turnover and anesthetic-sparing responses to dexmedetomidine. NE turnover in the hippocampus was significantly reduced following chronic administration of dexmedetomidine at both doses, and could be further lowered with acute administration of a sympatholytic dose (30 µg·kg^{-1}). The sympatholytic effect of acute dexmedetomidine was also present in the locus coeruleus (LC) of chronically treated rats. The baseline MAC and the MAC-sparing effect to acutely administered dexmedetomidine (30 µg·kg^{-1}) were preserved following chronic dexmedetomidine treatment (5 and 10 µg·kg^{-1}·hr^{-1}). In the EEDQ experiments, the decrease in NE turnover induced by dexmedetomidine required less than 20% and greater than 4% α_2 adrenoceptor availability in the LC. The MAC-sparing effect of dexmedetomidine was intact when 40% of α_2 adrenoceptors in the LC were available and blunted when receptor availability was reduced to 20%.

The distinctive refractory quality of the sympatholytic and MAC-sparing effects is explained by a comparatively more efficient signal transduction process with more receptor reserve than is needed for the hypnotic and analgesic responses. Thus, the MAC-sparing and NE turnover effects of α_2 adrenergic agonists persist after less robust responses have been attenuated by prolonged α_2 agonist administration.

THE QUEST FOR THE IDEAL α_2 ADRENERGIC AGONIST

Modern anesthetic drug discovery will likely be based on the concept that receptor subtypes exist and that designing drugs for one subtype can eliminate the side-effects mediated by a related subtype. Each of the clinically-efficacious agonists (clonidine, dexmedetomidine, mivazerol) have an imidazole ring which facilitates binding to non-adrenergic imidazoline sites. Also, the current agonists do not discriminate between the α_2 adrenoceptor subtypes. This relative non-specificity and non-selectivity may result in side-effects such as bradycardia and hypotension which is thought to be produced, in part, through the imidazoline receptor. Also, the acute hypertension that follows rapid bolus administration of α_2 agonists may be due to activation of the α_{2B} adrenoceptor, a receptor subtype which may not be involved in anesthetic or analgesic responses to α_2 agonists. In order to realize the full therapeutic potential of the α_2 agonists, a new generation of compounds needs to be developed which does not activate the imidazoline receptor or non-selectively activate non-anesthetic/analgesic α_2 receptor subtypes. This can be achieved by target-based drug design in which the spatial orientation of the receptor subtype is modeled.

DETERMINING THE RELEVANT ANESTHETIC/ANALGESIC RECEPTOR SUBTYPE

Antisense technology

Oligodeoxynucleotides (ODNs) of 15–30 nucleotides in length would have an unique sequence relative to the entire human genome; therefore antisense oligodeoxynucleotide (ODN) technology provides a degree of specificity which is lacking in conventional pharmacologic or toxicologic probes. Inhibition of gene expression by antisense ODNs relies on the ability of an ODN to bind a complementary sense mRNA sequence and prevent translation of the mRNA since the RNA strand of the RNA-DNA duplex becomes a substrate for the ubiquitous H RNAse. The level of the expressed protein will decrease and therefore the function propagated by the protein will be lacking. The effect on protein expression is usually incomplete unless the ODN is continuously delivered over a prolonged period of time ($\approx 2 \times$ turnover time). If the remaining protein exceeds the threshold required to produce a functional response, the decrement in protein expression will not be functionally noticed. Such "redundancy" does not exist for the behavioral responses to α_2 agonists, since the hypnotic response is lost when as few as 20% of the α_2 adrenoceptors in the LC are rendered dysfunctional. A robust feature of an *in vivo* antisense ODN study is that the effect should be reversible over time, the exact period of time being predicated by the turnover rate of the protein. The fact that the response is reversible assures one that the effect is not due to non-specific toxicity. This provides a degree of specificity which is lacking in "knockout" or "transgenic" experiments which are described below. There are now several examples in which antisense ODNs have been successfully used *in vivo* to interfere with specific protein synthesis and its physiologic function (Mizobe *et al.*, 1996b).

Gene targeted mice

Over the last 5 years several hundred mice strains have been developed in which the functioning of a specific gene has been knocked out or altered ("transgene") by gene targeting techniques. A vector containing the null mutation or the altered gene is introduced into the genome of mouse embryonic stem (ES) cells in tissue culture by homologous recombination between the targeted genomic locus and DNA introduced into the ES cells. A selectable marker permits identification of the successfully transfected ES cells which are heterozygous for the inactivated gene. These cells are injected into the recipient blastocysts for return to pseudo-pregnant foster mothers, where they can develop into chimeric pups. A few of the resulting chimeric progeny contain germline with ES cell contribution allowing transmission to future generations which are non chimeric heterozygous for the null allele. Heterozygotes are bred to produce animals homozygous for the inactivated gene. This technique provides a powerful tool for exploring the pharmacology of the behavior responses to α_2 agonists.

Preliminary studies with the D79N transgenic mouse in which coupling of the α_{2A} adrenoceptors to potassium and calcium ion channels is disrupted (Supranant *et al.*, 1992) reveal the involvement of this subtype and ion channels in the anesthetic, analgesic and sympatholytic action of α_2 agonists. Preliminary studies with α_{2B} adrenoceptor knockout mice reveal loss of the hypertensive response while the analgesic/anesthetic response remains intact (Link *et al.*, 1996).

DETERMINING THE 3-DIMENSIONAL STRUCTURE OF THE TARGET PROTEIN

Adrenoceptors belong to a class of proteins referred to as G protein-coupled receptors (GPCRs). The more than 400 members of the GPCR superfamily bind to structurally diverse ligands, ranging from cationic monoamines to large glycoproteins, to produce an equally diverse array of cellular actions. Despite this diversity, hydrophobicity plots on the amino acid sequences of each of the cloned GPCRs reveal a common pattern of seven hydrophobic sequences. The hydrophobic domains in each receptor are similar in size (20–28 amino acids) and of sufficient length to span the lipid membrane as α helices with intervening hydrophilic loops exposed intra- and extra-cellularly. Along with this topographic similarity there is greater than 20% sequence homology among the GPCRs, mostly in the transmembrane domains.

The tertiary structure of GPCRs has proven insoluble because of the difficulty in extracting large amounts of pure protein from the natural membranes for crystallography studies. In the absence of biophysical analysis, structural models have been devised, based largely on the folding pattern of the ancient retinal-linked visual pigment bacteriorhodopsin, one of only three integral membrane proteins for which three-dimensional information is known. Bacteriorhodopsin, a prokaryotic proton pump, is found in large amounts in naturally occurring lattices within the purple membrane of *Halobacterium halobium*. Although bacteriorhodopsin is not coupled to G proteins, it is similar to the opsins since both have a covalently linked retinal chromophore, which undergoes photo-isomerization to initiate either proton pump action (in the case of bacteriorhodopsin) or signal transduction (in the case of rhodopsin — see below). The first structural studies of bacteriorhodopsin (level of resolution = 5.0 Å) utilized electron-diffraction to determine the orientation of the transmembrane regions. Using this method investigators showed that bacteriorhodopsin has seven α helices arranged in a bundle, three of which are perpendicular to the plane of the lipid bilayer and four slightly tilted. This characteristic feature of seven transmembrane helices is used as a scaffold for molecular modeling of the GPCR. Subsequently, a higher resolution structural study, to a level of 3.0 Å, was achieved with electron cryo-microscopy resolution. This method refined the orientation of the helices.

Using a site-directed mutagenesis approach we now have concluded that human adrenoceptors exhibit the same helical packing arrangement as that found in bacteriorhodopsin (Mizobe *et al.*, 1996a). This provides the necessary validation for bacteriorhodopsin to be used as a template on which to model adrenoceptors. The molecular models can now be used as a target into which novel chemical entities can be "docked."

References

Correa-Sales, C., Rabin, B.C. and Maze, M. (1992a). A hypnotic response to dexmedetomidine, an α_2 agonist is mediated in the locus coeruleus in rats. *Anesthesiology*, **76**, 948–952.

Correa-Sales, C., Reid, K. and Maze, M. (1992b). Pertussis toxin-mediated ribosylation of G proteins blocks the hypnotic response to an α_2 agonist in the locus coeruleus of the rat. *Pharmacology, Biochemistry, Behavior*, **43**, 723–727.

Guo, T.-Z., Tinklenberg, J., Oliker, R. and Maze, M. (1991). Central α_1 adrenoceptor stimulation functionally antagonizes the hypnotic response to dexmedetomidine, an α_2 adrenoceptor agonist. *Anesthesiology*, **75**, 252–256.

Hayashi, Y., Guo, T.-Z. and Maze, M. (1995). Desensitization to the behavioral effects of α_2 adrenergic agonists in rats. *Anesthesiology*, **82**, 954–962.

Link, R.E., Stevens, M.S., Kulatunga, M., Scheinin, M., Barsh, G.S. and Kobilka, B.K. (1995). Targeted inactivation of the gene encoding the mouse α_{2C} adrenoceptor homolog. *Mol. Pharmacol.* (in press).

Maier, C., Steinberg, G.K., Sun, G.H., Tian-Zhi, G. and Maze, M. (1993). Neuroprotection by the α_2 adrenoceptor agonist, dexmedetomidine, in a focal model of cerebral ischemia. *Anesthesiology*, **79**, 306–12.

Mizobe, T., Maghsoudi, K., Sitwala, K., Tianzhi, G., Ou, J., Maze, M. (1996b). Antisense technology reveals the α_{2a}-adrenoceptor to be the subtype mediating the hypnotic response to the highly-selective agonist, dexmedetomidine in the rat. *J. Clinical Investigation*. In press.

Nacif-Coelho, C., Correa-Sales, C., Chang, L.L. and Maze, M. (1994). Perturbation of ion channel conductance alters the hypnotic response to the α_2 adrenergic agonist dexmedetomidine in the locus coeruleus of the rat. *Anesthesiology*, **81**, 1527–1534.

Rabin, B.C., Reid, K., Guo, T.-Z., Gustaffson, E. and Maze, M. (1996). The sympatholytic and MAC-sparing responses are preserved in rats rendered tolerant to the hypnotic and analgesic action of α_2 adrenergic agonists. *Anesthesiology* (under review).

Reid, K., Hayashi, Y., Guo, T.-Z., Correa-Sales, C., Nacif-Coelho, C. and Maze, M. (1994). Chronic administration of an α_2 adrenergic agonist desensitizes rats to the anesthetic effects of dexmedetomidine. *Pharmacology Biochemistry Behavior*, **47**, 171–175.

Reid, K., Hsu, J., Maguire, P.A., Rabin, B.C., Guo, T.-Z. and Maze, M. (1996). Chronic administration of dexmedetomidine decreases α_2-adrenergic receptor binding affinity, ribosylation of PTX-sensitive G proteins, and inhibition of adenylyl cyclase in the locus coeruleus of rats. *Pharmacol. Biochemistry and Behavior* (under review).

Segal, I.S., Vickery, R.G., Sheridan, B.C., Doze, V.A. and Maze, M. (1988). Dexmedetomidine diminishes halothane anesthetic requirements in rats through a postsynaptic α_2 adrenergic receptor. *Anesthesiology*, **69**, 818–823.

Suprenant, A., Horstman, D.A., Akbarali, H. and Limbird, L.E. (1992). Point mutation of the α_2-adrenoceptor that blocks coupling to potassium but not calcium currents. *Science*, **257**, 977–980.

Vickery, R.G., Sheridan, B.C., Segal, I.S. and Maze, M. (1988). Anesthetic and hemodynamic effects of the stereoisomers of medetomidine, an α_2 adrenergic agonist, in halothane-anesthetized dogs. *Anesth. Analg.*, **67**, 611–615.

Zornow, M.H., Maze, M., Dyck, B. and Shafer, S.L. (1993). Dexmedetomidine decreases cerebral blood flow velocity in humans. *J. Cerebral Blood Flow and Metabolism*, **13**, 350–3.

α_2 AND α_1-ADRENERGIC RECEPTORS IN THE REGULATION OF PERIPHERAL VASCULAR FUNCTION

MICHAEL T. PIASCIK,[*][‡] MARTA S. SMITH,[‡] STEPHANIE E. EDELMANN,[‡] LEIGH B. MACMILLAN,[§] and LEE E. LIMBIRD[§]

[‡]*The Department of Pharmacology and the Vascular Biology Research Group, University of Kentucky, College of Medicine, Lexington, Kentucky, USA and* [§]*The Department of Pharmacology, Vanderbilt University, Nashville, Tennessee, USA*

Specific aspects of the regulation of peripheral vascular function by subtypes of the α_2-adrenergic receptor (AR) have been studied and compared to that observed with α_1-AR subtypes. To evaluate the contribution of the α_{2A}-AR a genetically modified mouse line was developed by homologous recombination. The mutation, D79Nα_{2A}AR, substitutes an asparagine for an aspartate at position 79 of the amino acid sequence of the receptor. Resting mean arterial blood pressure (MAP) or heart rate in D79N mutant animals was not different when compared to wild-type. In wild-type animals, infusion of UK 14304 produced a dose-dependent increase in MAP which was followed by hypotension which lasted for 3 hrs. In contrast, UK 14304 had no pressor effect in D79N mutants nor was the long lasting hypotension observed. In the isolated wild-type and D79N aorta, UK 14304 produced a concentration dependent increase in contractile tension which was antagonized by prazosin. However, the maximal response in mutant aorta was significantly less than that seen in the wild-type. These data suggest the α_{2A}-AR plays a prominent role in peripheral as well as central cardiovascular regulation.

Parallels exist in the regulatory patterns seen with the α_1-AR. In specific blood vessels a single subtype appears to be the predominant regulatory receptor. Experiments with the α_{1D}-selective antagonist BMY 7378 and the use of antisense oligonucleotides suggest that the α_{1D}-AR is the predominant contractile receptor in the aorta, femoral and iliac arteries. The α_{1A}-AR appears to be the predominant receptor regulating contraction in the caudal, mesenteric resistance and renal arteries.

KEY WORDS: α-adrenergic receptor subtypes, antisense oligonucleotides, gene substitution, homologous recombination, peripheral vascular regulation

INTRODUCTION

Peripheral and central α_2 and α_1-ARs are utilized by the sympathetic nervous system to regulate systemic arterial blood pressure and blood flow (Bylund *et al.*, 1995; Ruffolo *et al.*, 1993). It was originally suggested that the α_1-ARs are associated with the neuroeffector junction and respond to sympathetic nerve stimulation while the α_2-ARs exist extraneuronally (Langer *et al.*, 1981; Langer and Shepperson, 1982). More recently Faber and associates (Faber, 1988; McGillivray-Anderson and Faber, 1991) showed that in the skeletal muscle microcirculation the α_1-ARs regulate larger arterioles (80–120 microns)

Correspondence to: Dr Michael Piascik, Dept. of Pharmacology, University of Kentucky, AB Chandler Medical Center MN-305, Lexington KY 40536-0001, USA. Tel: 606-323-5107, Fax: 606-323-1981

while the α_2-AR appears to be the major regulatory receptor in smaller diameter (< 80 microns) arterioles. Furthermore, both subtypes are innervated and mediate vasoconstriction following activation of the sympathetic nervous system (Ohyanagi et al., 1991).

While subtypes of both the α_2 and α_1-AR have been shown to exist (Bylund et al., 1995; Graham et al., 1996; Ruffolo et al., 1993), the exact contribution of these subtypes to the regulation of vascular smooth muscle contraction is not known. Ping and Faber (1993) showed that mRNA for the α_{2A}-AR is located in aortic smooth muscle. We have shown that mRNA for all three α_1-ARs is co-localized in a variety of smooth muscles (Piascik et al., 1995; Guarino et al., 1995). In this report we have used complementary genetic, molecular biological and pharmacologic approaches to gain a better understanding of the contribution of subtypes of both the α_2 and α_1-AR families to vascular regulation.

METHODS

Generation of a mouse line with a mutant α_{2A}-AR gene

A genetically-modified mouse line was developed by homologous recombination. The target of the mutation was an aspartic acid at position 79 in the α_{2A}-AR, a residue predicted to lie within the second transmembrane region of the α_{2A}-AR and highly conserved among G-protein coupled receptors. Hit and run targeting of the α_{2A}-AR gene resulted in the successful substitution of an asparagine for aspartate 79 in the mouse genome (Limbird et al., 1994). Generation of animals homozygous for this mutation was confirmed by diagnostic restriction digests, Southern analysis and PCR analysis of the DNA isolated from homozygous offspring.

Hemodynamic recordings

Wild-type and D79N mutant mice were anesthetized with a mixture of ketamine and acepromazine. The femoral vein was cannulated for intravenous drug administration and the femoral artery cannulated to record systemic arterial blood pressure and heart rate. The mouse was allowed to recover fully from the anesthetic. Increasing amounts of UK 14304 were injected and the effect on blood pressure and heart rate noted. Blood pressure and heart rate were monitored for 4 hours after the last dose (100 µg/kg) of UK 14304.

In vitro assessment of contractile function

Arteries were removed from wild-type and D79N mutant mice as well as male Sprague-Dawley rats and placed in a cold physiological saline solution (PSS) of the following composition: NaCl, 130 mM; KCl, 4.7 mM; KH_2PO_4, 1.18 mM; $MgSO_4$-7 H_2O, 1.17 mM; $CaCl_2$-2 H_2O, 1.6 mM; $NaHCO_3$, 14.9 mM; dextrose, 5.5 mM; Na_2 EDTA, 0.03 mM. Contractile responses to UK 14304 or phenylephrine were determined as we have previously described (Piascik et al., 1995).

Introduction of antisense α_1-AR oligonucleotides into the vasculature

A 30% w/v solution of pluronic F-127 gel (Sigma) containing 150 µM sense or antisense oligonucleotides directed against the translational start site of the α_{1D}-AR mRNA was pre-

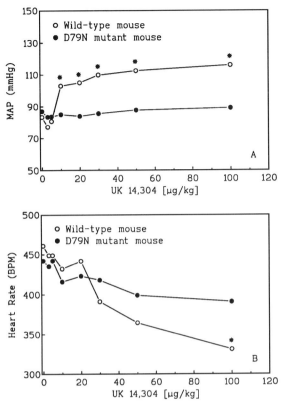

FIGURE 1 Acute effects of increasing amounts of UK 14304 on MAP and HR in wild-type and D79N mutant mice. * indicates statistical significance as determined by ANOVA.

pared in water at 4°C. Sequences of these constructs were: Sense-GAGATGACTTTCCGA, Antisense-GTCTCGGAAAGTCAT. All pipet tips and storage tubes were kept at 4°C. Following a surgical incision, the femoral artery was located and gently cleaned of adhering tissue. The artery was gently lifted and a silicon sleeve placed around the vessel and held in place with a loosely tied suture. 20 µl of the pluronic gel/oligonucleotide solution was injected on the artery and held in place with the sleeve. The gel hardened and solidified around the artery. The suture was tightened to hold the sleeve in place but not so tight as to restrict blood flow. The artery was gently placed back into its original position and the wound closed with surgical staples.

RESULTS

α_{2A}-ARs

The resting mean arterial blood pressure (MAP) and heart rate (HR) were not different in the D79N mutant mice (89 ± 3 mmHg and 442 ± 38 bpm) when compared to the wild-type (87 ± 2 mmHg and 461 ± 36 bpm). Injection of UK 14304 produced a transient, concentration dependent and statistically significant increase in MAP in wild-type animals. Associated with this hypertension was a decrease in HR (Figures 1A and 1B). The

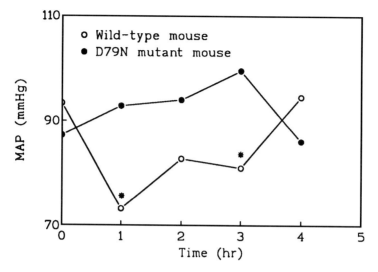

FIGURE 2 Long term effects of UK 14304 on MAP in wild-type and mutant mice. * indicates statistical significance as determined by ANOVA.

hypertensive response to UK 14304 was not observed in the D79N mutant (Figure 1). Following the last dose (100 µg/kg) of UK 14304 into the wild-type mouse there was a sustained decrease in MAP which was significant for 3 hours (Figure 2). This sustained hypotension was not observed in the D79N mutant (Figure 2).

In aorta isolated from wild-type mice, UK 14304 induced a concentration dependent increase in contractile tension. This response was significantly reduced in the mice harboring the D79N mutation in the α_{2A}-AR (Figure 3).

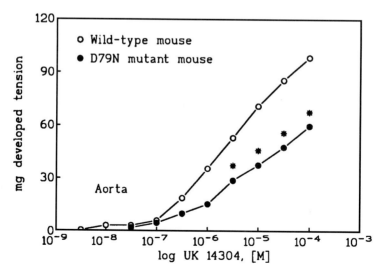

FIGURE 3 Effects of increasing amounts of UK 14304 on the response of the aorta isolated from wild-type and mutant mice. * indicates statistical significance as determined by ANOVA.

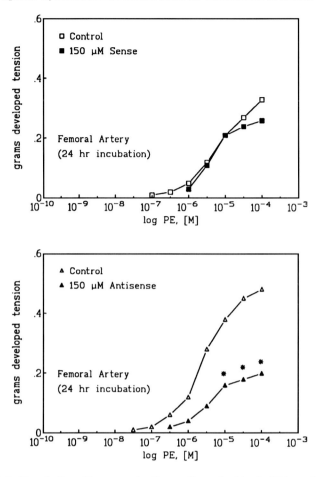

FIGURE 4 Effect of transfection with sense and antisense oligonucleotides for 24 hr on the response of the femoral artery to phenylephrine. * indicates statistical significance as determined by ANOVA.

α_1-ARs

Transfection with an α_{1D}-AR antisense oligonucleotide significantly blocked the response of the femoral artery to phenylephrine. In contrast a sense oligonucleotide had no effect on the phenylephrine response (Figure 4).

DISCUSSION

α_{2A}-ARs

Homologous recombination was used to introduce a mutant α_{2A}-AR substituting an asparagine for aspartic acid at position 79 into the mouse genome. The amino acid is predicted to lie within the second transmembrane span of the α_{2A}-AR and is thought to be vital for coupling to G-proteins (Ceresa and Limbird, 1994). When this mutation was expressed

Shown are the mean and standard deviation of independent α_{1A}, α_{1B} and α_{1D}-AR ribonuclease protection assays. Assays were performed on 30 μg of total RNA at different times, with different RNA preparations and [^{35}S]-labeled probes.

TABLE 1
Estimated picograms of mRNA for the α_1-ARs in peripheral blood vessels.

	α_{1A}-AR[a] (n = 2)	α_{1B}-AR[a] (n = 3)	α_{1D}-AR[b] (n = 3)
Aorta	797 ± 478	134 ± 86	113 ± 70
Caudal	2052 ± 859	161 ± 122	31 ± 17
Iliac	1358 ± 628	226 ± 104	25 ± 24
Mesenteric	2317 ± 680	130 ± 25	20 ± 11
Renal	2788 ± 297	159 ± 29	88 ± 36

[a]Data from Guarino et al., 1996.
[b]Data from Piascik et al., 1995.

in vitro in cultured AtT20 pituitary cells, the mutated receptor was selectively uncoupled from activation of K$^+$ currents but retained the ability to block Ca^{2+} currents and adenylyl cyclase (Surprenant et al., 1992).

Resting MAP or HR were not affected in mutant mice. This implies that the α_{2A}-AR does not function in the tonic maintenance of blood pressure. However, other receptors could be upregulated and substitute for the α_{2A}-AR in maintaining resting pressure. The ability of UK 14304 to increase MAP was completely blocked in the D79N mutant mouse indicating that peripheral α_{2A}-ARs associated with blood vessels play a prominent role in mediating the pressor response to exogenously administered agonists.

The long lasting hypotensive response to UK 14304 was also completely blocked in the D79N mutant mouse. This hypotension is produced by receptors in the central nervous system. These data strongly implicate central α_{2A}-ARs in mediating hypotension to exogenously administered agonists, even with imidazoline structures. This counters current thinking that the central effects of clonidine for example are due to presynaptic "imidazoline" receptors rather than the α_2-AR.

The spasmogenic actions of UK 14304 were significantly reduced in aorta isolated from D79N animals. This would suggest that, in mutant animals, the D79N α_{2A}-AR was poorly coupled to the second messenger pathways leading to contraction. In preliminary studies we have shown that the response in the D79N mutant aorta is blocked by 0.3 and 1 μM prazosin. In these concentrations, prazosin does not bind to either the wild-type or D79N mutant α_{2A}-AR. Therefore, the response to UK 14304 in the mutant aorta is mediated by either or both the α_{2B} and α_{2C}-AR. Nonetheless, the response to UK 14304 is decreased in the mutant animal, suggesting that the response of this artery reflects a combination of α_2-AR inputs. Mutation removes the contribution of the α_{2A}-AR and results in an alteration in UK 14304 response.

α_1-ARs

Recent evidence has shown that mRNA for all α_1-ARs can be detected in peripheral blood vessels. An example of this can be seen in Table 1 which is a compilation of our previous work. The mRNA for the α_{1A}-AR is expressed in much higher levels, being approximately 90% of the total α_1-AR mRNA. In contrast, radioligand binding studies have consistently estimated the expression of α_{1A}-AR protein to be a small percentage (5–36%) of the total α_1-AR population (Piascik et al., 1990; Morrow and Creese, 1986; Hanft and Gross, 1989). Therefore, in the α_1-AR family there does not seem to be a correlation between mRNA and protein expression.

In the α_1-AR family, α_{1D}-AR mRNA is the rarest. In the aorta and iliac artery, low concentrations of the α_{1D}-AR selective antagonist BMY 7378 block the contractile response of phenylephrine (Piascik et al., 1995). These same concentrations had little effect on the phenylephrine response in the caudal, mesenteric resistance and renal arteries (Piascik et al., 1995). Therefore, despite being only a small portion of the mRNA population, the α_{1D}-AR appears to be the major α_1-AR regulating the contraction of the aorta and iliac artery.

We transfected α_{1D}-AR sense and antisense oligonucleotides into the femoral artery. When these oligonucleotides were fluorescently labelled we detected a significant fluorescent signal in the femoral artery 24 hr after transfection. Furthermore, we show that neither sense nor antisense oligonucleotides impaired the response to KCl or 5-HT. In contrast, antisense impaired the contractile response to phenylephrine. These data suggest that the α_{1D}-AR is also involved in regulating femoral artery contraction.

Work by Han et al. (1990) has suggested that the α_{1A} is the major α_1-AR regulating the function of the caudal, mesenteric resistance and renal arteries. Our recent work (Piascik et al., 1995) would support this conclusion. Therefore, it appears that, of the α_1-AR subtype family, a contractile role for only the α_{1A} and α_{1D}-ARs has been established. Nonetheless we do know that mRNA for α_{1B}-AR is present in all peripheral arteries. Evidence is emerging which indicates this receptor may play a role in cell growth and proliferation. Therefore, the contractile regulating receptors could be the α_{1A} and α_{1D}-ARs with the α_{1B} subserving a poorly understood growth regulating function.

In summary, we have used complementary experimental strategies to reveal the diversity and complexity of cardiovascular regulation by the α_2 and α_1-AR families.

References

Bylund, D.B., Regan, J.W., Faber, J.E., Hieble, J.P., Triggle, C.R. and Ruffolo, R.R. (1995). Vascular α-adrenoceptors: from the gene to the human. Cand. J. Physio. Pharmacol., **73**, 533–543.

Ceresa, B.P. and Limbird, L.E. (1994). Mutation of an aspartate residue highly conserved amoung G protein coupled receptors results in non-reciprocal disruption of α_2-adrenergic receptor-G protein interactions. J. Biol. Chem., **269**, 29557–29564.

Faber, J.E. (1988). In situ analysis of α-adrenoceptors on arteriolar and venular smooth muscle in the rat skeletal muscle microcirculation. Circ. Res., **62**, 37–50.

Graham, R.M., Perez, D.M., Hwa, J. and Piascik, M.T. (1996). α_1-adrenergic receptor subtypes: molecular structure, function and signalling. Circ. Res., in press.

Guarino, R.D., Perez, D.M. and Piascik, M.T. (1996). Recent advances in the molecular pharmacology of the α_1-adrenergic receptors. Cellular Signalling, **8**, in press.

Han, C., Li, J. and Minneman, K.P. (1990). Subtypes of α_1-adrenoceptors in rat blood vessels. Eur. J. Pharmacol., **190**, 97–104.

Hanft, G. and Gross, G. (1989). Subclassification of α_1-adrenoceptor recognition sites by urapidil derivatives and other selective antagonists. Br. J. Pharmacol., **97**, 691–700.

Langer, S.Z. and Shepperson, N.B. (1982). Recent developments in vascular smooth muscle pharmacology: The postsynaptic α_2-adrenoceptor. Trends in Pharmacol. Sci., **3**, 440–444.

Langer, S.Z., Shepperson, N.B. and Massingham, R.B. (1981). Preferential noradrenergic innervation of the α-adrenergic receptors in vascular smooth muscle. Hypertension, **Suppl. 3**, I-112–I-118.

Limbird, L.E., MacMillan L.B. and Keefer, J.R. (1995). Specificity in α_2-adrenoceptor signal transduction: Receptor subtypes, coupling to distinct signal transduction pathways and localization to discrete subdomains in target cells. Pharmacol. Comm., **6**, 139–145.

McGillivray-Anderson, K.M. and Faber, J.E. (1991). Effect of reduced blood flow on α_1 and α_2-adrenoceptor constriction of rat skeletal muscle microvessels. *Circ. Res.*, **69**, 165–173.

Morrow, A.L. and Creese, I. (1986). Characterization of α_1-adrenergic receptor subtypes in rat brain: a reevaluation of [^{3}H]WB4104 and [^{3}H]prazosin binding. *Mol. Pharmacol.*, **29**, 321–330.

Ohyanagi, M., Faber, J.E. and Hishigoki, K. (1991). Differential activation of α_1 and α_2-adrenoceptors on microvascular smooth muscle during sympathetic nerve stimulation. *Circ. Res.*, **68**, 233–244.

Piascik, M.T., Guarino, R.D., Smith, M.S., Soltis, E.E., Saussy, D.L. and Perez, D.M. (1995). The specific contribution of the novel α_{1D}-adrenoceptor to the contraction of vascular smooth muscle. *J. Pharmacol. Expt. Ther.*, **275**, in press.

Piascik, M.T., Butler, B.T., Pruitt, T.A. and Kusiak, J.W. (1990). Agonist interaction with alkylation sensitive and resistant α_1-adrenoceptor subtypes. *J. Pharmacol. Exp. Ther.*, **254**, 982–991.

Ping, P. and Faber, J.E. (1993). Characterization of the α-adrenoceptor gene expression in arterial and venous smooth muscle. *Am. J. Physiol.*, **265**, H1501–H1509.

Ruffolo, R.R. Jr., Nichols, A.J., Stadel, J.M. and Heible, J.P. (1993). Pharmacologic and therapeutic applications of α_2-adrenoceptor subtypes. *Ann. Rev. Pharmacol and Tox.*, **32**, 243–279.

Surprenant, A., Horstman, D.A., Akbarali, H. and Limbird, L.E. (1992). A point mutation of the α_2-adrenoceptor that blocks coupling to potassium but not calcium currents. *Science*, **257**, 977–980.

THERAPEUTIC USE OF α_2-ADRENOCEPTOR AGONISTS IN GLAUCOMA

J. BURKE,[*] C. MANLAPAZ,[*] A. KHARLAMB,[*] E. RUNDE,[*]
E. PADILLO,[*] C. SPADA,[*] A. NIEVES, S. MUNK,[†] T. MACDONALD,[‡]
M. GARST,[†] A. ROSENTHAL,[§] A. BATOOSINGH,[§] R. DAVID,[§]
J. WALT[§] and L. WHEELER[*]

*Departments of [*]Biological Sciences, [†]Chemical Sciences, and [§]Clinical Research;
Allergan, Inc., Irvine, CA 92715, USA*
[‡]*Department of Chemistry, University of Virginia, Charlottesville, VA 22901, USA*

The glaucomas are a heterogeneous group of ocular disorders in which there is progressive damage to the optic nerve resulting in reduced visual field and eventual blindness. Current therapies focus on reducing intraocular pressure (IOP), one of the primary risk factors. α_2-Adrenoceptor agonists are a potent class of compounds that lower IOP but widespread use has been limited because of side-effects. One strategy for reducing these side-effects has been the development of compounds that are more selective for the α_2- over the α_1-adrenoceptor. The result of this strategy is the clinical development of brimonidine which has α_2-adrenoceptor selectivity superior to that of clonidine or *p*-aminoclonidine, two agonists that also lower IOP in humans. A one-month dose-response study in ocular hypertensive and glaucoma patients showed that the 0.2% concentration was safe and effective. In two controlled multi-centered studies, brimonidine 0.2% was found to be safe and effective in decreasing IOP for up to six and 12 months duration. The next step in the evolution of α_2-adrenoceptor agonists for glaucoma is the evaluation of additional patient benefits with compounds that show selectivity for subtypes of the α_2-adrenoceptor.

KEY WORDS: Brimonidine, glaucoma, IOP, mydriasis, vasoconstriction, α_2-adrenoceptor

INTRODUCTION

Glaucoma is a family of sight-threatening disorders in which there is progressive damage to the optic nerve resulting in loss of vision and eventual blindness. The pathophysiological mechanism(s) of optic nerve damage is unknown but increased intraocular pressure (IOP) and cupping of the disk are important risk factors often associated with primary open angle glaucoma, the most common form in the US and Europe. Glaucoma is estimated to affect more than two million Americans and is the leading cause of blindness in the US. In primary open angle glaucoma, aqueous humor can freely flow through the chamber angle between the iris and cornea of the eye. Narrow angle or closed angle glaucoma is often treated with surgery (Shields, 1992b).

The current strategy for the treatment of primary open angle glaucoma is to lower IOP which presumably decreases the mechanical stress on the optic nerve. IOP is a balance

---

Correspondence to: Dr. Larry Wheeler, Allergan Inc., 2525 DuPont Drive, Irvine CA 92715-1599, USA. Tel: 714-752-450, Fax: 714-246-5578

between aqueous humor production and the resistance to the outflow of fluid from the eye. Interest in α_2-adrenoceptor agonists was stimulated by reports in the late 1970s that topically applied clonidine decreased IOP (Harrison and Kaufmann, 1977). The lack of success of clonidine as a glaucoma treatment in the US has been the observation of significant systemic side-effects such as sedation and cardiovascular depression. Progress was made in this area with the clinical testing of *p*-aminoclonidine, an analog of clonidine, where decreases in IOP were observed without significant systemic side effects (Yuksel *et al.*, 1992). *p*-Aminoclonidine is currently marketed in the US for acute uses such as the prevention of IOP spikes following anterior segment laser surgery (Robin, 1990) and to manage patients with uncontrolled IOP receiving maximally tolerated medical therapy (Lish *et al.*, 1992). *p*-Aminoclonidine is a potent vasoconstrictor that produces conjunctival hypoxia (Serdahl *et al.*, 1989) and anterior segment vasoconstriction (Chandler and DeSantis, 1985; Fahrenbach *et al.*, 1989) following topical administration. The concern that chronic ocular blood flow reduction may lead to adverse consequences has led to continuing research and clinical evaluation of more α_2-adrenoceptor agonists. Brimonidine (UK-14,304) is the next ophthalmic α_2-adrenoceptor agonist to be evaluated in clinical trials. It has been evaluated in multi-center clinical studies and showed efficacy in controlling IOP spikes associated with argon laser trabeculoplasty (Barnebey *et al.*, 1993; David *et al.*, 1993), and in providing long-term IOP control in glaucoma and ocular hypertensive patients (Schuman, 1995). The focus of this chapter will be on comparing the pre-clinical and clinical pharmacology of three ophthalmic α_2-adrenoceptor agonists in order to describe the utility of these agents in treating glaucoma and discuss the prospects of this class of agents to continue to meet the medical needs of the glaucoma patient.

RESULTS AND DISCUSSION

In vitro and in situ pharmacology

The α-adrenoceptor pharmacology of brimonidine, clonidine and *p*-aminoclonidine were compared in radioligand binding and tissue bath bioassays. Affinity to human α_1- and α_2-adrenoceptors were assessed by [^{3}H]prazosin and [^{3}H]rauwolscine binding in human cerebral cortex and the human colonic cell line (HT-29), respectively. α_1-Adrenoceptor activation was measured by the contractile response of the isolated rabbit iris dilator muscle. The iris dilator muscle contains an α_1-adrenoceptor which mediates a contractile response (Konno and Takayanagi, 1986) that results in increased pupil size or mydriasis, an undesirable effect in glaucoma therapy. Ocular α_1-adrenoceptor stimulation is also associated with ciliary vasoconstriction (Fahrenbach *et al.*, 1989, 1991) which reduces blood flow. Activation of α_2-adrenoceptors was determined by inhibition of electrically-induced contractions of the isolated rabbit vas deferens. The results (Table 1) show that all three compounds had a higher affinity for α_2- than α_1-adrenoceptors. Brimonidine was 30 times more α_2-adrenoceptor selective than *p*-aminoclonidine and 15 times more selective than clonidine. The intrinsic activity of brimonidine and clonidine for contracting the iris dilator muscle were lower than that of *p*-aminoclonidine. These results suggest that of the three α_2-adrenoceptor agonists, brimonidine would be the least likely to produce α_1-adrenoceptor-mediated side effects such as mydriasis and ocular vasoconstriction.

TABLE 1
Receptor pharmacology of brimonidine, p-aminoclonidine and clonidine at α_1- and α_2-adrenoceptors. These data have been previously presented (Burke et al., 1995b; Burke et al., 1995c; Wheeler et al., 1995).

Compound	Radioligand Binding K_i (nM)[a]		Functional EC_{50} (nM)[a]		Selectivity[f]
	α_1[b]	α_2[c]	α_1[d]	α_2[e]	
Brimonidine	1850 ± 322 (5)	1.9 ± 0.5 (6)	2650 ± 341 (8)	1.0 ± 0.1 (24)	974,2650
Clonidine	513 ± 108 (4)	3.4 ± 0.4 (6)	943 ± 163 (6)	4.4 ± 0.4 (11)	67,214
p-Aminoclonidine	181 ± 18 (4)	6.0 ± 0.9 (5)	216 ± 16 (14)	1.9 ± 0.2 (9)	30,113

[a] Mean ± SEM; 'N' is noted in parentheses.
[b] [^{3}H]Prazosin binding in human cerebral cortex adapted from protocol for rat cerebral cortex as described by (Reader et al., 1987). K_d = 0.27 ± 0.01 nM; B_{max} = 272 ± 11 fmoles/mg protein (n = 4).
[c] [^{3}H]Rauwolscine binding in HT-29 cells as described by (Bylund et al., 1988). K_d = 0.33 ± 0.04 nM; B_{max} = 136 ± 5.9 fmoles/mg protein (n = 2).
[d] Contraction of isolated rabbit iris dilator muscle as described by (Konno and Takayanagi, 1986). Intrinsic activity relative to 100 μM (−) norepinephrine for brimonidine, clonidine and p-aminoclonidine were 0.3, 0.3 and 0.5, respectively.
[e] Inhibition of electrically-induced contractions in the rabbit vas deferens as described by (Lattimer and Rhodes, 1985).
[f] Ratio (α_1/α_2).

To assess the potential for human ocular vasoconstriction, the three α_2-adrenoceptor agonists were examined in the microvasculature of human retinal tissue transplanted in the hamster cheek pouch (Spada et al., 1995). Drugs were administered by localized topical microsuffusion to the abluminal side of the arteriolar segment associated with the microvasculature of the retinal xenograft. Arteriolar caliber was assessed by intravital microscopy. Brimonidine was not vasoactive up to 10 μM. A concentration of 100 μM decreased vessel caliber ~ 8%. Clonidine evoked a marked concentration-dependent decrease in arteriolar caliber of up to 35% over the 10 nM to 100 μM concentration range. p-Aminoclondine was the most potent compound in this model, producing a 21% decrease in caliber with concentrations as low as 0.01 nM. These data suggest a correlation between affinity at α_1-adrenoceptors and inducing vasoconstriction in the human retinal microvasculature.

In vivo pharmacology

The ocular responses to topically administered brimonidine, clonidine and p-aminoclonidine were compared in animal models. Figure 1 shows the peak ocular hypotensive (Panel A) and pupil diameter (Panel B) changes in normotensive rabbits. The ocular hypotensive response to brimonidine was concentration-related with a maximum response of 5.1 mm Hg. Clonidine and p-aminoclonidine were only marginally effective in the normotensive rabbit. This lack of ocular hypotensive efficacy is related, in part, to the unique sensitivity of this species to α_1-adrenoceptor stimulation. α_1-Adrenoceptor stimulation increases IOP in rabbits; α_2-adrenoceptor stimulation decreases IOP (Burke et al., 1995d; Burke and Potter, 1986; Innemee et al., 1981; Murray and Leopold, 1985). Thus, IOP at any given time is a composite of two opposing dynamics. The higher the α_2-receptor selectivity, the greater the ocular hypotensive effect. p-Aminoclonidine has higher affinity for α_1-adrenoceptors and consequently has less of an ocular hypotensive response. The other

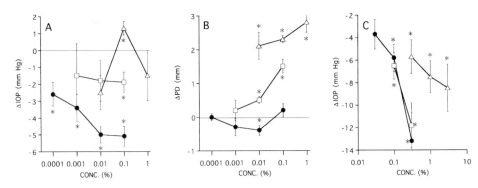

FIGURE 1 Dose response curves for the ocular responses to topically administered brimonidine (●), clonidine (□) and p-aminoclonidine (△) in conscious New Zealand White (albino) rabbits and ocular hypertensive cynomolgus monkeys. Panel A: IOP changes in ocularly normotensive rabbits; Panel B: Pupil diameter changes in rabbits; Panel C: IOP changes in monkeys made ocularly hypertensive by argon laser trabecularplasty of the trabecular meshwork as described by (Sawyer and McGuigan, 1988). Drugs were applied topically and unilaterally as a single 50 μl eye-drop. Intraocular pressure and pupil diameter were measured non-invasively with a 30R model Digilab pneumatonometer and a mm ruler, respectively. Twenty-five μl of an anesthetic (proparacaine) was topically applied before IOP measurements to minimize ocular discomfort due to tonometry. Two baseline measurements were made prior to instillation of the drugs, followed by periodic measurements up to 6 hours post-instillation. Each data point represents mean change from pre-treatment levels ± standard error of the mean, and represents at least 6 animals. Asterisks indicate a significant difference from saline control; $p < 0.05$, unpaired Student's t-test. These data have been previously presented (Burke et al., 1995b; Burke et al., 1995c; Wheeler et al., 1995).

consequence of α_1-adrenoceptor stimulation in rabbits is mydriasis. The low α_1-receptor affinity of brimonidine is manifested in the absence of a mydriatic response at concentrations that lower IOP. In contrast, p-aminoclonidine at equivalent concentrations elicits a profound and prolonged mydriatic response (> 6 hours). It should be noted that the rabbit overstates the case for the mydriatic potential of imidazolines such as p-aminoclonidine. The profound mydriasis produced by p-aminoclonidine in rabbits was smaller in monkeys (Gabelt et al., 1994) and of variable magnitude in humans (Abrams et al., 1987; Robin, 1988).

The ocular hypotensive activity of topically administered brimonidine, clonidine and p-aminoclonidine were evaluated in an experimental model of ocular hypertension in monkeys in which the trabecular meshwork was partially occluded by an argon laser (Figure 1, Panel C). This model is widely regarded as one which closely approximates glaucoma in humans since the clinical hallmarks of glaucoma develop from the ocular hypertension (Podos et al., 1987). The three agonists produced a concentration-dependent decrease in IOP with a ranked order of potency of brimonidine = clonidine > p-aminoclonidine (p-aminoclonidine was ~10-fold less potent). Brimonidine and clonidine were also more efficacious than p-aminoclonidine. The maximum ocular hypotensive response to brimonidine in this study, 13 mm Hg, was less than the 20 mm Hg previously reported for the same concentration in ocular hypertensive monkeys (Serle et al., 1991b). The ocular hypotensive effect in monkeys was without evidence of an ocular hypertensive component. This IOP profile contrasts with the rabbit and underlines a species difference of the IOP response to α_2-adrenoceptor agonists. Another species difference is the receptor(s) which mediates the ocular hypotensive response. In rabbits, ocular hypotension is

TABLE 2

The effects of unilateral topical administration of brimonidine, the non-selective β-blocker, timolol, and the $β_1$-selective blocker, betaxolol, on aqueous humor flow and IOP in cynomolgus monkeys. Aqueous humor flow rates were measured fluorophotometrically in ketamine-sedated ocularly normotensive monkeys as described by (Robinson and Kaufman, 1993) and averaged from hourly scans for 4 hours following drug instillation. IOP was determined in conscious ocularly normotensive and hypertensive monkeys using a 30R model Digilab pneumatonometer. The peak IOP responses within 4 hours of drug administration are reported. Drugs were applied as a single 50 µl drop. Pre-treatment values for aqueous humor flow rate, normotensive IOP and hypertensive IOP averaged 1.5 µl/min, 15 mm Hg and 28 mm Hg, respectively.

		% Decrease	
		Intraocular Pressure (IOP)	
Compounds	Aqueous Flow	Normotensive	Hypertensive
Saline	1.3 ± 5 (7)[a]	3.4 ± 5 (6)	5 ± 3 (6)
0.3% Brimonidine	40 ± 5 (6)[b]	43 ± 3 (6)[b]	40 ± 4 (6)[b]
0.5% Timolol	42 ± 8 (6)[b]	9 ± 3 (6)	44 ± 5 (6)[b]
0.5% Betaxolol	32 ± 3 (6)[b]	6.2 ± 4 (6)	33 ± 5 (6)[b,c]

[a] Mean ± standard error of the mean with number on monkeys in parentheses.
[b] Significant difference comparing drug treated with saline treated animals using an unpaired Student's t-test, $p < 0.05$.
[c] Concentration of betaxolol in ocular hypertensive monkeys was 1%, 25 µl.

mediated by peripheral $α_2$-adrenoceptors (Burke et al., 1989); in monkeys, an imidazoline receptor appears to play a significant role since idazoxan is more efficacious than rauwolscine at blocking the ocular hypotensive response to brimonidine. The vascular postjunctional $α_2$-adrenoceptor does not play a role in the IOP response to brimonidine in rabbits and monkeys because of insensitivity to the selective antagonists SKF 104078, SKF 105854 and SKF 104191 (Burke et al., 1995a).

An important mechanism of action of $α_2$-adrenoceptor agonists is the suppression of aqueous humor inflow. Clonidine has been shown to decrease inflow in humans (Lee et al., 1984), but not in monkeys (Bill and Heilmann, 1975). p-Aminoclonidine decreases inflow in rabbits (Bartels, 1989), monkeys (Gabelt et al., 1994) and humans (Gharagozloo et al., 1988; Toris et al., 1995b). Brimonidine decreases flow in monkeys (Gabelt et al., 1994; Serle et al., 1991b) and humans (Toris et al., 1995a). Ophthalmic β-blockers also act by inhibiting aqueous humor production. Table 2 compares the flow and IOP effects of brimonidine, the non-selective β-blocker timolol, and the $β_1$-selective blocker betaxolol, in cynomolgus monkeys. The responses on aqueous flow suppression and IOP reduction in ocular hypertensive monkeys were comparable. However, unlike the β-blockers, brimonidine lowered IOP equally well in normotensive eyes suggesting that brimonidine may have other mechanisms of action in addition to aqueous humor suppression.

Other mechanisms for reducing IOP include an increase in the outflow of aqueous humor. There are two types of outflow mechanisms: trabecular and uveoscleral. Trabecular or conventional flow is via the trabecular meshwork and the canalicular system. Uveoscleral outflow describes the exit of aqueous from the eye via non-trabecular routes, principally the root of the ciliary muscle and into the suprachoroidal space (Shields, 1992a). The outflow enhancing effects of some $α_2$-adrenoceptor agonists have been described. Brimonidine increases uveoscleral outflow in rabbits (Lee et al., 1992; Serle et al., 1991a) and humans (Toris et al., 1995a). p-Aminoclonidine increases trabecular outflow in

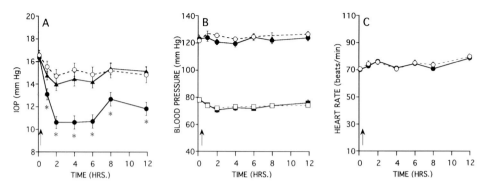

FIGURE 2 The IOP (Panel A), blood pressure (Panel B) and heart rate (Panel C) responses to unilateral application of 0.5% brimonidine (salt) in normal human volunteers. Filled and open symbols represent the drug-treated and vehicle-treated groups, respectively. The IOP response in the contralateral, untreated eye is represented by (▲) in Panel A. Systolic and diastolic blood pressures in Panel B are represented by the upper and lower curves, respectively. The number of subjects in each treatment group was 23. Asterisks represent a statistical difference from vehicle treated eyes as determined by an unpaired Student's t-test. These data have been previously presented (David et al., 1995).

humans (Toris et al., 1995b). Brimonidine appears to affect inflow and uveoscleral outflow equally in humans. This combination of aqueous humor flow suppression and uveoscleral outflow enhancing appears to be unique to brimonidine.

Clinical pharmacology

The clinical use of the three ophthalmic α_2-adrenoceptor agonists, clonidine, p-aminoclonidine and brimonidine has been recently reviewed by (Chacko and Camras, 1994) and (Serle, 1994). All three agonists effectively decrease IOP in glaucoma patients. Unlike clonidine, p-aminoclonidine and brimonidine treatments are not associated with clinically-significant systemic side effects such as sedation and hypotension. The cardiovascular response to brimonidine in humans is different from monkeys. The ocular hypotensive response to brimonidine in monkeys is associated with a symmetric contralateral IOP response and concomitant decreases in blood pressure and heart rate, which suggests CNS mediation (Burke et al., 1995a; Gabelt et al., 1994; Serle et al., 1991b). In humans, brimonidine can decrease IOP without a contralateral IOP response, hypotension or bradycardia (Figure 2). Additionally, in a one-month dose-ranging study (Derick et al., 1993) with 168 glaucoma or ocular hypertensive patients randomized to three brimonidine concentrations (0.08%, 0.2%, 0.5%) and vehicle, and another controlled multi-center study (Schuman, 1995) of 12 months duration, involving over 200 ocular hypertensive or glaucoma subjects treated with the 0.2% concentration, brimonidine decreased IOP without clinically significant sedative or cardiovascular changes.

The 12 month brimonidine clinical trial also revealed that ocular side effects (burning, stinging, hyperemia, foreign body sensation, blurring) were comparable to Timoptic® and within acceptable limits. The incidence of ocular allergy (conjunctivitis, blepharoconjunctivitis, follicular conjunctivitis) was low, less than 10%. Some allergy is associated with most new ophthalmic medications. The incidence of ocular allergy with

p-aminoclonidine is reported to be as high as 45–48% (Butler et al., 1995; Robin, 1995). The difference in allergy rates between brimonidine and p-aminoclonidine may be explained by the structural differences between the two molecules. p-Aminoclonidine has substantial structural similarity to amodiaquine, which would suggest that a quinone diimine species (with its potent electrophilicity, allergenic and necrotic potential), could be an intermediate in oxidative metabolism (Sanner and Higgins, 1991). The oxidative potential of brimonidine is higher than p-aminoclonidine which has an oxidation potential that is similar to amodiaquine (Munk et al., 1996). Thus, brimonidine is expected to have reduced potential as a sensitizer.

α_2-Adrenoceptor agonists have a place in the present and future armamentarium of glaucoma medications. Currently, this compound class is the most effective for the treatment of the rise in IOP that is associated with anterior segment laser surgery. α_2-Adrenoceptor agonists are useful as adjuncts to existing medical therapies and are effective in patients on maximally tolerated medical therapy. Some α_2-adrenoceptor agonists (e.g. brimonidine) have good potential for use as firstline mono therapy. The future of α_2-adrenoceptor agonists to treat glaucoma may lie in the identification of agonists that are selective for subtypes of the α_2-adrenoceptor. Subtype selective agonists may further improve the therapeutic index of this class of compounds. Additionally, recent studies have shown that α_2-adrenoceptor agonists are neuroprotective agents. α_2-adrenoceptor agonists reduce infarct size in models of focal ischemia (Ruffolo et al., 1995), upregulate neuronal survival factors such as bFGF in the retina (Wen et al., 1995) and prevent photoreceptor degeneration from excess light exposure (Wen et al., 1996). Brimonidine was shown to facilitate rescue and slow down degeneration of optic nerve fibers that escaped primary injury following a partial crush to the rat optic nerve (Yoles et al., 1996). Perhaps the development of subtype selective agonists or agonists that offer protection to the optic nerve will provide new approaches for the treatment of glaucoma and provide additional patient benefits.

References

Abrams, D.A., Robin, A.L. and Pollack, I.P. (1987). The safety and efficacy of topical 1% ALO 2145 (p-aminoclonidine hydrochloride) in normal volunteers. *Arch. Ophthalmol.*, **105**, 1205–1207.

Barnebey, H.S., Robin, A.L., Zimmerman, T.J., Morrison, J.C., Hersh, S.B., Lewis, R.A., et al. (1993). The efficacy of brimonidine in decreasing elevations in intraocular pressure after laser trabeculoplasty. *Ophthalmology*, **100**, 1083–1088.

Bartels, S.P. (1989). Effects of apraclonidine (p-aminoclonidine) on aqueous humor dynamics and pupil size. *Invest. Ophthalmol. Vis. Sci., (Suppl)*, **30**, 371.

Bill, A. and Heilmann, K. (1975). Ocular effects of clonidine in cats and monkeys (macaca irus). *Exp. Eye Res.*, **21**, 481–488.

Burke, J., Crosson, C. and Potter, D. (1989). Can UK-14,304-18 lower IOP in rabbits by a peripheral mechanism? *Curr. Eye Res.*, **8**, 547–552.

Burke, J., Kharlamb, A., Shan, T., Runde, E., Padillo, E., Manlapaz, C., et al. (1995a). Adrenergic and imidazoline receptor-mediated responses to UK-14,304-18 (brimonidine) in rabbits and monkeys: A species difference. In *The Imidazoline Receptor: Pharmacology, Functions, Ligands, and Relevance to Biology and Medicine*, edited by D. Reis, P. Bousquet, and A. Parini, pp. 78–95. New York: The New York Academy of Sciences.

Burke, J., Manlapaz, C., Padillo, E., Kharlamb, A., Runde, E., Nieves, A., et al. (1995b). Receptor and ocular pharmacology of the α_2-adrenoceptor agonists, brimonidine, p-aminoclondine and clonidine. *International Symposium on Experimental and Clinical Ocular Pharmacology and Pharmaceutics*, Geneva, Switzerland, 29.

Burke, J., Manlapaz, C., Padillo, E., Runde, E. and Spada, C. (1995c). α-Adrenoceptor and ocular pharmacology of brimonidine, p-aminoclonidine and clonidine. *XX Pan-American Congress of Ophthalmology*, Quito, Equador, 110.

Burke, J., Padillo, E., Manlapaz, C. and Wheeler, L. (1995d). The effect of vascular post-junctional α_2-adrenoceptor antagonists on the intraocular pressure (IOP) response to brimonidine in rabbits. *The Faseb Journal, (Suppl)*, **9**, A106.

Burke, J.A. and Potter, D.E. (1986). Ocular effects of a relatively selective α_2 agonist (UK-14,304-18) in cats, rabbits and monkeys. *Curr. Eye Res.*, **5**, 665–676.

Butler, P., Mannschreck, M., Lin, S., Hwang, I. and Alvarado, J. (1995). Clinical experience with the long-term use of 1% apraclonidine: Incidence of allergic reactions. *Arch. Ophthalmol.*, **113**, 293–296.

Bylund, D.B., Ray-Prenger, C. and Murphy, T.J. (1988). α_{2A} and α_{2B} adrenergic receptor subtypes: antagonist binding in tissues and cell lines containing only one subtype. *J. Pharmacol. Exp. Ther.*, **245**, 600–607.

Chacko, D.M. and Camras, C.B. (1994). The potential of α_2-adrenergic agonists in the medical treatment of glaucoma. *Curr. Opinion Ophthalmol.*, **5**, 76–84.

Chandler, M. and DeSantis, L. (1985). Studies of *p*-aminoclonidine as a potential antiglaucoma agent. *Invest. Ophthalmol. Vis. Sci., (Suppl)*, **26**, 227.

David, R., Spaeth, G.L., Clevenger, C.E., Perell, H.F., Siegel, L.I., Henry, J.C., et al. (1993). Brimonidine in the prevention of intraocular pressure elevation following argon laser trabeculoplasty. *Arch. Ophthalmol.*, **111**, 1387–1390.

David, R., Walters, T.R., Sargent, J.B., Batoosingh, A. and Walt, J. (1995). The safety and hypotensive efficacy of brimonidine tartrate 0.08%, 0.2%, 0.35%, and 0.5% in normotensive subjects. *Eur. J. Ophthalmol., (Suppl)*, **5**, 156.

Derick, R.J., Walters, T.R., Robin, A.L., Barnebey, H.S., Choplin, N.T., Kelley, E.P., et al. (1993). Brimonidine tartrate: a one month dose response study. *Invest. Ophthalmol. Vis. Sci., (Suppl)*, **34**, 929.

Fahrenbach, W.H., Bacon, D.R. and Van Buskirk, E.M. (1989). Vasoactive drug effects on the uveal vasculature of the rabbit a corrosion casting study. *Invest. Ophthalmol. Vis. Sci., (Suppl)*, **30**, 100.

Fahrenbach, W.H., Bacon, D.R. and Van Buskirk, E.M. (1991). Microvascular responses to chronic glaucoma drug therapy. *Invest. Ophthalmol. Vis. Sci., (Suppl)*, **32**, 944.

Gabelt, B.T., Robinson, J.C., Hubbard, W.C., Peterson, C.M., Debink, N., Wadhwa, A., et al. (1994). Apraclonidine and brimonidine effects on anterior ocular and cardiovascular physiology on normal and sympathectomized monkeys. *Exp. Eye Res.*, **59**, 633–644.

Gharagozloo, N.Z., Relf, S.J. and Brubaker, R.F. (1988). Aqueous flow is reduced by the α-adrenergic agonist, apraclonidine hydrochloride (ALO 2145). *Ophthalmology*, **95**, 1217–1220.

Harrison, R. and Kaufmann, C.S. (1977). Clonidine: Effects of a topically administered solution on intraocular pressure and blood pressure in open-angle glaucoma. *Arch. Ophthalmol.*, **95**, 1368–1373.

Innemee, H.C., de Jonge, A., van Meel, J.C.A., Timmermans, P.B.M.W.M. and van Zwieten, P.A. (1981). The effect of selective α_1 and α_2-adrenoceptor stimulation on intraocular pressure in the conscious rabbit. *Arch. Pharmacol.*, **316**, 294–298.

Konno, F. and Takayanagi, I. (1986). Characterization of postsynaptic α_1 adrenoceptors in the rabbit dilator smooth muscle. *Arch. Pharmacol.*, **333**, 271–276.

Lattimer, N. and Rhodes, K.F. (1985). A difference in the affinity of some selective α_2-adrenoceptor antagonists when compared on isolated vasa deferentia of rat and rabbit. *Naunyn-Schmiedeberg's Arch. Pharmacol.*, **329**, 278–281.

Lee, D.A., Topper, J.E. and Brubaker, R.F. (1984). Effect of clonidine on aqueous humor flow in normal human eyes. *Exp. Eye Res.*, **38**, 239–246.

Lee, P.-Y., Serle, J.B., Podos, S.M. and Severin, C. (1992). Time course of the effect of UK 14304-18 (brimonidine tartrate) on rabbit uveoscleral outflow. *Invest. Ophthalmol. Vis. Sci., (Suppl)*, **33**, 1118.

Lish, A.J., Camras, C.B. and Podos, S.M. (1992). Effect of apraclonidine on intraocular pressure in glaucoma patients receiving maximally tolerated medications. *J. Glaucoma*, **1**, 19–22.

Munk, S.A., Wiese, A., Thompson, C.D. and Macdonald, T. (1996). Oxidation potential and allergic response of α_2-agonists. *Invest. Ophthalmol. Vis. Sci., (Suppl)*, **37**, S832.

Murray, D.L. and Leopold, I.H. (1985). α-adrenergic receptors in rabbit eyes. *J. Ocular Pharmacol.*, **1**, 3–18.

Podos, S., Camras, C., Serle, J. and Lee, P.-Y. (1987). Pharmacological alteration of aqueous humor dynamics in normotensive and glaucomatous monkeys eyes. In *Glaucoma Update III*, edited by G.K. Krieglstein, pp. 225–235. Berlin: Springer-Verlag.

Reader, T.A., Briere, R. and Grondin, L. (1987). α_1 and α_2 adrenoceptor binding in cerebral cortex: Competition studies with [^{3}H]prazosin and [^{3}H]idazoxan. *J. Neural Transm.*, **68**, 79–95.

Robin, A.L. (1988). Short-term effects of unilateral 1% apraclonidine therapy. *Arch. Ophthalmol.*, **106**, 912–915.

Robin, A.L. (1990). The role of apraclonidine hydrochloride in laser therapy for glaucoma. *Trans. Am. Ophthalmol. Soc.*, **87**, 729–761.

Robin, A.L. (1995). Questions concerning the role of apraclonidine in the management of glaucoma. *Arch. Ophthalmol.*, **113**, 712–714.

Robinson, J.C. and Kaufman, P.L. (1993). Dose-dependent suppression of aqueous humor formation by timolol in the cynomolgus monkey. *J. Glaucoma*, **2**, 251–256.

Ruffolo, R.R. Jr., Bondinell, W. and Hieble, J.P. (1995). α- and β-Adrenoceptors: From the gene to the clinic. 2. Structure-activity relationships and therapeutic applications. *J. Med. Chem.*, **38**, 3681–3716.

Sanner, M.A. and Higgins, T.J. (1991). Chemical basis for immune mediated idiosyncratic drug hypersensitivity. In *Annual Reports in Medicinal Chemistry*, edited by J.A. Bristol, pp. 181–190. New York: Academic Press.

Sawyer, W.K. and McGuigan, L.J. (1988). Effects of antiglaucoma agents on the unanesthetized, trained primate in both the ocular normotensive and ocular hypertensive state. *Invest. Ophthalmol. Vis. Sci., (Suppl)*, **29**, 81.

Schuman, J. (1995). The long term safety and efficacy of brimonidine 0.2% in the treatment of glaucoma and ocular hypertension. *International Symposium on Experimental and Clinical Ocular Pharmacology and Pharmaceutics*, Geneva, Switzerland, 29.

Serdahl, C.L., Galustian, J. and Lewis, R.A. (1989). The effects of apraclonidine on conjunctival oxygen tension. *Arch. Ophthalmol.*, **107**, 1777–1779.

Serle, J. (1994). Pharmacological advances in the treatment of glaucoma. *Drugs and Aging*, **5**, 156–170.

Serle, J., Podos, S., Lee, P.-Y., Camras, C. and Severin, C. (1991a). Effects of α_2-adrenergic agonists on uveoscleral outflow in rabbits. *Invest. Ophthalmol. Vis. Sci., (Suppl)*, **32**, 867.

Serle, J.B., Steidl, S., Wang R.F., Mittag, T.W. and Podos, S.M. (1991b). Selective α_2-adrenergic agonists B-HT 920 and UK14,304-18: Effects on aqueous humor dynamics in monkeys. *Arch. Ophthalmol.*, **109**, 1158–1162.

Shields, M.B. (1992a). Aqueous humor dynamics I: Anatomy and physiology. In *Textbook of Glaucoma*, 3rd edn, pp. 5–14. Baltimore: Williams and Williams.

Shields, M.B. (1992b). An overview of glaucoma. In *Textbook of Glaucoma*, 3rd edn, pp. 1–2. Baltimore: Williams and Williams.

Spada, C.S., Nieves, A.L., Burke, J.A. and Woodward, D.F. (1995). Comparative effects of brimonidine, *p*-aminoclonidine and clonidine on arteriolar caliber in human retinal tissue. *Invest. Ophthalmol. Vis. Sci., (Suppl)*, **36**, 1041.

Toris, C.B., Gleason, M., Camras, C.B. and Yablonski, M.E. (1995a). Effects of brimonidine on aqueous humor dynamics in human eyes. *Arch. Ophthalmol.*, in press.

Toris, C.B., Tafoya, M.E., Camras, C.B. and Yablonski, M.E. (1995b). Effects of apraclonidine on aqueous humor dynamics in human eyes. *Ophthalmol.*, **102**, 456–461.

Wen, R., Cheng, T., Li, Y. and Steinberg, R.H. (1995). Induction of bFGF gene expression *in vivo* in rat photoreceptors by the α_2-adrenergic agonists xylazine and clonidine. *Soc. Neurosci. Abstr., (Suppl)*, **21**, Part 2, 1045.

Wen, R., Cheng, T., Li, Y. and Steinberg, R.H. (1996). Systemic application of α_2-adrenergic agonists xylazine or clonidine induces bFGF expression in photoreceptors. *Invest. Ophthalmol. Vis. Sci., (Suppl)*, **37**, S436.

Wheeler, L., Manlapaz, C., Padillo, E., Kharlamb, A., Runde, E., Neives, A., *et al.* (1995). A comparison of the receptor and ocular pharmacology of the α_2-adrenoceptor agonists, brimonidine, *p*-aminoclonidine and clonidine. *Eur. J. Ophthalmol., (Suppl)*, **5**, 156.

Yoles, E., Muler, S., Schwartz, M., Burke, J., WoldeMussie, E. and Wheeler, L. (1996). Injury-induced secondary degeneration of rat optic nerve can be attenuated by α_2-adrenoceptor agonists AGN 191103 and brimondine. *Invest. Ophthalmol. Vis. Sci., (Suppl)*, S114.

Yuksel, N., Guler, C., Caglar, Y. and Elibol, O. (1992). Apraclonidine and clonidine: a comparison of efficacy and side effects in normal ocular hypertensive volunteers. *Int. Ophthalmol.*, **16**, 337–342.

SUBJECT INDEX

(Page numbers followed by 'f' indicate figures, those followed by 't' indicate tables)

AGN 190851 5
agonist independent activity of receptors 77–83
agonist-promoted desensitization of α_2C10-adrenergic receptor 114–115
α_2-adrenergic receptor agonists in the perioperative period 161–169
α_2-adrenergic receptors
  acylation 124
  adenylyl cyclase 86, 106t
  blood pressure 2–4, 163, 168, 176
  intracellular calcium 86–91
  glycosylation 124
  MAP kinase 97, 98t
  metabolic labeling 126
  mood disorders 149–159
  p21ras 97–98
  phosphoinositide-3 kinase 98–99, 98t
  phospholipase C 86, 91
  phospholipase D 97
  phosphorylation 114
  tyrosine kinases 95–100
α_2-adrenergic receptor agonists
  adjuncts in anesthesia induction 165–168
  agonist binding 58
α_{2A}-adrenergic receptor 6, 7, 8t, 9, 10, 21f, 45, 46, 55, 86, 89, 90, 96, 97, 105, 124–126
  binding buffer 109
  C10 11
  D79N mutants 99–100, 168, 171–176
  GT1 neuronal cells 131–140
  hippocampal neurons 131–140
  human 8t
  immunoreactivity 129–140, 142–148, 144f, 145f
  porcine 8t
  regulation 105, 108, 113–118, 134, 142, 147
  subcellular location 136–138, 142–148
α_{2B}-adrenergic receptor 6–10, 8t, 86, 87, 89, 105, 124–126
  binding buffer 108
  C2 11
  human 8t
  immunoreactivity 142–148, 144f, 145f
  mouse 8t
  rat 8t
  regulation 105, 108, 113–118, 142, 147
  subcellular distribution 142–148
α_{2C}-adrenergic receptor 6–10, 89, 105, 124–126, 142
  C4 11
  binding buffer 108
  human 8t
  immunoreactivity 142–148
  mouse 8t
  opossum 8t
  rat 8t
  regulation 105, 113–118
  subcellular distribution 142–148
α_{2D}-adrenergic receptor 6–10, 87
  binding buffer 109
  mouse 9t
  rat 9t
α_{1B}-adrenergic receptor 79, 107, 176t, 177
α_{1C}-adrenergic receptor 176t, 177
α_{1D}-adrenergic receptor 176t, 177
allosteric model of receptor action 78–79, 78f
allosteric-ternary complex model of receptor action 78–79, 78f
alprenolol 44
analgesia: effects of α_2-adrenergic receptor agonists 162
anesthesia: effects of α_2-adrenergic receptor agonists 162–163

antagonist-mediated changes in fluorescence from IANBD-labeled β₂-adrenergic receptor antibodies
   α_{2A}-adrenergic receptor 129–140, 141–148
   α_{2B}-adrenergic receptor 141–148
   α_{2C}-adrenergic receptor 141–148
   human transferrin receptor 131, 133
   Igp120 131, 133
   kinesin H1 131, 137
   mannose-6-phosphate receptor 131, 133
antisense oligonucleotides 168
   use in intact artery 172–173, 175–176, 175f, 176t
anxiolysis: effects of α_2-adrenergic receptor agonists 165–168
ARC-239 7, 10
astrocytes 89

BAM1303 104
basal activity of receptor 81–82
β_2-adrenergic receptor 32, 33, 44, 45, 50, 79, 81
B-HT 920 2, 5
B-HT 933 2, 5
biotinylation of α_2-adrenergic receptor subtypes 126
bovine
   aortic smooth muscle cells 146, 147f
   pineal 7, 9t, 10
   retina 9t, 10
BRL 44408 7

canine
   adipocyte 8t
   saphenous artery 3
cardiovascular system: effects of α_2-adrenergic receptor ligands 2–4, 163, 176
caveolae 63
   purification 67
   signalling hypothesis 68f
caveolin 64
   acylation 69, 70f, 72
   interaction with G-protein 71, 73, 73f, 74
   interaction with receptor 69, 70f
   isoforms 69, 73, 73f

   expression 66
   membrane topology 65f, 66
   oligomerization 70f, 71
   phosphorylation 66
   transformation 67
cell lines transfected with adrenergic receptors
   CHO 46, 89, 104f, 105t, 105, 108, 109f, 113
   CHW-1102 114
   COS 45, 113, 119, 143f, 144f, 145f
   HEK293 113
   JEG-3 86
   MDCK 124–126
   NIH-3T3 57, 105, 105t, 106f, 109f
   PC-12 57, 87
   Rat-1 97–100
   S115 86
   sf9 cells 33, 86
central nervous system: effects of α_2-adrenergic receptor ligands 3, 149–159, 162–163
CHO cell line 46, 89, 104f, 105, 105f, 108, 109f, 113
CHW-1102 cell line 114
clonidine 1, 4, 152, 161–163, 180–185, 182f
constitutive activity of receptor 22t, 41, 79, 81
COS cell line 45, 113, 119, 143f, 144f, 145f
cysteine probes 33

D79N mutant of α_{2A}-adrenergic receptor 99–100, 168, 171–176
depression: noradrenergic system 150–151
depression: treatment with α_2-adrenergic receptor antagonists 155–160, 156t
desensitization of α_2-adrenergic receptor subtypes 113–121, 115f
dexmedetomidine 87, 87f, 90f, 143, 166–167
down-regulation of α_2-adrenergic receptor subtypes 105–107

Easson–Stedman Hypothesis 43–44, 47, 49
EEDQ (1-ethoxycarbonyl-2-ethoxy-1,2-dihydroquinoline) 107
enantiomers of catecholamines 47, 49, 50

endocrine system: effects of α_2-adrenergic
    receptor ligands 3, 164
epitope tags 33, 125–127

fluorescent labeling of the β_2-adrenergic
    receptor 31–40
fluorescence spectroscopy 31, 32
fluorescence properties of IANBD-labeled
    β_2-adrenergic receptor 34f

gain of function mutations in receptor
    activation 79–81
gastrointestinal system: effects of α_2-adrenergic
    receptor ligands 3, 164–165
glutathione sepharose 55, 58
gel overlay 55
glaucomas
    α-adrenergic receptor pharmacology *in vitro*
        180–181
    α-adrenergic receptor pharmacology *in vivo*
        181–184
    α_2-adrenergic receptors — clinical
        pharmacology 184–185
    use of α_2-adrenergic receptors agonists
        179–187
guinea pig
    atria 10
    brain cortex 10
    ileum 10
guanine nucleotide binding proteins 19
    activator 58, 60
    β subunit 24
    γ subunit 24
    GTPase activity 21, 22
    GTPγ ^{35}S binding 55–57, 56f, 57f
    photoaffinity labeling 24
    receptor interaction 23f, 25f
GT1 cell line 130

hamster adipocyte 9t
HEL cells 89
HEK293 cell line 113
hippocampal neurons 130
human
    adipocyte 8t
    kidney 10
    ocular ciliary body 143
    platelet 8t, 151–152
    saphenous vein 10
HT-29 6, 8t, 10, 105, 106f, 107

idazoxan 6, 163
    as antidepressant 154–155, 157f, 158f
imidazolines
    Easson Stedman Hypothesis 44
    receptors 152, 176
immunolocalization of native α_2-adrenergic
    receptor subtypes 141–147, 146f, 147f
insulin secretion 3
intracellular α_{2A}-adrenergic receptor subtypes
    in neurons and GT1 cell line 129–139,
    132f, 133f, 134f, 135f, 136f, 137f
intracellular domains of α_{2A}-adrenergic
    receptor and proposed function 23f
intraocular pressure 4, 163
    betaxolol 183, 183t
    brimonidine (UK14304) 180–185, 182f,
        183t, 184f
    clonidine 1, 4, 180–185, 182f
    p-aminoclonidine 180–185, 182f
    timilol 183, 183t
internalization of α_2-adrenergic receptor
    subtypes 144f, 145f
inverse agonists 38–41, 78–80
isoproterenol 35f, 44

JEG-3 cell line 86

ligand recognition properties of α_2-adrenergic
    receptor subtypes 8t, 9t
ligand-specific structural changes in
    adrenergic receptors 31–42
lipolysis 3
lysophosphatidic acid signalling 96, 98t, 99t

mastoparan 20
mechanism of agonist-induced receptor
    downregulation 107–108
mechanisms of inhibitory responses initiated
    by α_2-adrenergic receptors 85–86
mechanisms of receptor and G-protein
    coupling 19–28

mechanisms of stimulatory responses initiated by α_2-adrenergic receptors 86–91
methylnorepinephrine 2
Madin-Darby canine kidney cell line (MDCK cells) 123–126
melanin granule dispersion 3
microtubules and receptor localization in neurons 134f
MK912 6, 7
models of receptor conformation relative to active states 39, 40, 77, 78, 83
molecular biology of α_2-adrenergic receptors 11–12
mood disorders and α_2-adrenergic receptors 153t
multi-state model of receptor action 40f

naphazoline 1
negative antagonists 36f, 37–41, 78–79
neurotransmitter release 3
neutral antagonists 36f, 38, 41
NIH-3T3 fibroblasts 57, 105 106f, 109f
NG-10815 cell line 6, 8t, 10, 58
noradrenergic system in depression 151

opossum kidney 7, 8t
OK cell line 8t, 10, 106f
oxymetazoline 1, 8t

partial agonism 36f, 38–41
PC-12 cell line 57, 87
phenoxybenzamine 2
phentolamine 6
phenylephrine 2, 175, 175f
phosphorylation of α_2-adrenergic receptor subtypes 115–118, 116f, 117f, 119f
platelet α_2-adrenergic receptors in depression 151–152, 152t
postjunctional α_2-adrenergic receptors 2
prazosin 8t
precoupling of receptor and G-protein 78
prejunctional autoreceptors 1

radioligand binding assays 4
  receptor subclassification 6
rat-1 fibroblasts 97–100

rat
  aortic smooth muscle cells 143
  brain cortex 7, 10
  embryonic spinal cord 146
  enterocyte 9t, 10
  kidney 6, 8t, 10
  neonatal lung 8t
  spinal cord 143
  submaxillary gland 7, 9t, 10
  tail artery 3
  vas deferens 10
rabbit
  atria 10
  adipocyte 9t
  brain cortex 10
  heart 1
  iris dilator muscle 181t
  kidney 10
  ocular ciliary body 143, 146f
  pulmonary artery 1, 10
  saphenous veins 3
  vas deferens 180
receptor kinases 113, 118–120
  selective phosphorylation of α_2-adrenergic receptor subtypes 118–120, 119f
receptor repopulation kinetics 107
receptor trafficking 123–126
  basolateral surface of epithelial cells 125
regulation of α_2-adrenergic receptor subtypes by buffers 108–111
RINm5F cell line 9t, 10
RX821002 6, 10, 54, 109f, 110f, 111

sedation: effects of α_2-adrenergic receptors agonists 162, 165–168
SF-9 insect cell line 33, 86
S115 cell line 86
sedation 3, 165–168
signal transfer from receptor to G-protein 53–61
SK&F 104078 104
SK&F 104856 7
SK&F 35886 5
ST-91 5
stereoselective interactions of catecholamines with α_{2A}-adrenergic receptor 43–50
  effect of site-directed mutagenesis 48t

structure/activity of α_2-adrenergic receptor ligands 6
structural determinants of receptor activation 82–83
subclassification α_2-adrenergic receptors 6–12

ternary complex model of receptor action 78–79, 78f
third intracellular loop of α_2-adrenergic receptors 22, 23, 58
   constitutive activation 22, 79, 80
   peptides 20, 55
   phosphorylation 114, 114f, 117, 118
tools for pharmacological characterization of α_2-adrenergic receptors 5–6
tolerance to hypnotic and analgesic effects of α_2-adrenergic receptor agonists 166–167

trafficking of α_2-adrenergic receptor subtypes in epithelial cells 123–127, 125f
transmembrane helices of adrenergic receptors 44, 45, 49, 83
two-state model of receptor action 40f

UK-14304 5, 11, 54, 88f, 89, 97, 98, 98t, 99t, 171, 173–174, 173f, 174f, 176

WB 4101 4
WD-40 repeat in G-protein β subunit 24, 25f, 26

xylazine 1

Y-79 cell line 8t

ASPET COLLOQUIA

1994　Multiple Phosphodiesterases
　　　The Role of Adhesion Molecules in Cardiovascular Pharmacology

1995　Structure and Function of P_2-Purinoceptors
　　　Pharmacologic Interventions in Thrombosis and Thrombolysis
　　　Receptor-Acting Xenobiotics and their Risk Assessment
　　　α_2-Adrenergic Receptors: Structure, Function and Therapeutic Implications

1996　Effects of Gonadal Steroids on Vascular Function